THE TIME DILATION PARADOX

Navigating the Curious Phenomenon of Time Dilation

Ayush Malviya

SandalWood

ISBN-13: 9798851616129

Cover design by: Art Painter
Library of Congress Control Number: 2018675309
Printed in the United States of America

To All Physics enthusiast On Planet Earth

Time is a paradox, constantly distorting itself in response to our limited understanding of it. Time dilation takes us on a fascinating adventure in which the passage of time stretches and contracts, reshaping the very fabric of our lives. Come with us as we discover the mysteries of the cosmos by navigating the twisting passageways of time dilation.

CONTENTS

INTRODUCTION

As the sun dipped below the horizon, casting a warm golden glow over the city, a group of friends gathered on the rooftop of an apartment building. Laughter filled the air as they reminisced about their shared adventures and misadventures. Among them was Maya, a bright and curious student with a passion for science.

Maya's eyes twinkled with excitement as she began to share a story that had captured her imagination. She spoke of a renowned scientist named Dr. Johnson, who had devoted his life to unraveling the mysteries of time. It was said that he had discovered a phenomenon known as time dilation - a mind-bending concept that challenged the very fabric of our understanding.

Time dilation, Maya explained, was the idea that time could stretch or shrink depending on the relative speed of an object or its proximity to a gravitational force. It was as if time itself were elastic, capable of bending and warping in the presence of these powerful influences.

Intrigued, her friends leaned in closer, their eyes wide with wonder. Maya continued her tale, painting a vivid picture of Dr. Johnson's groundbreaking experiments. She described how he had synchronized atomic clocks, one on Earth and another aboard a spacecraft hurtling through space at incredible speeds. When the spacecraft returned to Earth, the clock on board was found to be slightly behind the one that had remained on the planet.

Time, Maya proclaimed, was not the rigid constant we had always believed it to be. It could be manipulated and distorted, a revelation that shook the very foundations of our understanding of the universe.

As the story unfolded, Maya's friends became captivated by the possibilities that time dilation presented. They imagined traveling to distant galaxies, where time would pass at a different pace, and returning to Earth to find that centuries had elapsed while mere years had gone by for them.

But Maya's tale was not just a flight of fancy; it was firmly grounded in the realms of science. She explained the principles of special relativity, which described how time dilation occurred as objects approached the speed of light. She delved into the intricacies of general relativity, where the curvature of space and time caused by massive objects led to gravitational time dilation.

The rooftop was abuzz with discussion as Maya's friends grappled with these mind-bending concepts. They marveled at the idea that time, a fundamental element of our existence, could be so malleable. Their curiosity ignited, they eagerly asked Maya for more details, hungry to understand the inner workings of this extraordinary phenomenon.

And so, dear reader, join me on a journey through the enigmatic realm of time dilation. Together, we will unravel its mysteries, exploring the intricate tapestry of special and general relativity, and diving deep into the gravitational forces that shape our universe. Prepare to be amazed, for the secrets of time await us, ready to challenge our perception of reality and expand the boundaries of our understanding.

THE CONCEPT OF TIME

We can't imagine making sense of life without the concept of time, which is intrinsic to how we take in and make sense of the world. Time is a broad and intricate term that includes duration, the flow of events, and the awareness of the past, the present, and the future.

Simply put, time is a sequence of events that unfolds in a linear fashion from the past to the present to the future. It provides structure for making sense of the world and making sense of our experiences within it. As our minds continually process and make sense of the ongoing sequence of events, we experience time through them.

Time's inherent irreversibility is one of its defining features. There is no way to go back in time and change what has already happened. The passage of time creates a feeling of transience because one moment always gives way to the next. Time also incorporates the idea of duration, which is the time frame in which something occurs or lasts. Seconds, minutes, hours, days, and years are all ways that we quantify and compare the passage of time.

Alteration and the passage of time go hand in hand. Things happen, conditions change, and things change as time goes on. Because time is always changing, we always feel like we're making headway in this world. With the passage of time, we may watch and learn about changes and progress. Without the passage of time, everything would appear to be fixed and unchanging.

The relative and subjective nature of time is also crucial. Individuals and their environments can color their perceptions of time in unique ways. Several things affect how we experience time, including our mood, the focus of our attention, and the depth of our emotions and experiences. Time is subjective, as seen by the adage "Time flies when you're having fun," which refers to the way in which pleasant events make time seem to pass quickly and unpleasant ones make it seem to move slowly.

Furthermore, memory is deeply intertwined with the idea of time. Our sense of time is shaped by our memories and our ability to plan for the future. Remembering things from the past gives our lives structure and consistency. They help us make sense of who we are by providing context for how we might place the events of the present within the context of history.

Time is also studied scientifically, most notably in physics. In his theory of relativity, Albert Einstein proposed the idea of time dilation, which argues that gravity and velocity can affect the passage of time. Depending on their location and velocity, viewers in different gravitational fields may experience time passing at different rates. As so, it underlines the complex nature of the interaction between time and place and undermines our intuitive sense of time as a universal constant.

Time, as we have seen, is an essential part of our lives because it gives us a way to structure and make sense of the events that occur to us. Time is all about evolution, change, and remembering the past. Time is relative; different people will see it and react to it in different ways. It's a broad idea with implications in fields as diverse as philosophy, psychology, physics, and more. Our sense of ourselves and the story of our lives is shaped by our encounters with the world and by the way we conceptualize and experience time.

THE DISCOVERY OF TIME DILATION

Time dilation challenged our preconceptions about how time works and how it relates to distance. This revolutionary idea arose as a direct result of Albert Einstein's theory of relativity, which radically altered our understanding of the nature of time, space, and the rules of the universe.

Time dilation is a theoretical idea that originated with Einstein's special relativity theory in 1905. According to the theory of special relativity, the physical laws do not change when moving from one observer to another. This theory proposed the idea that time can vary depending on the relative motion of observers, which casts doubt on the traditional view of time as an absolute and universal entity.

Special relativity relies heavily on the idea of "relative simultaneity." Events that happen at the same time to one observer may occur at different times to another observer who is moving with respect to the first. The implication of this theory is that people in different reference frames may perceive time differently.

The concept of "time dilation," wherein moving objects or people appear to experience a slower passage of time compared to a stationary observer, was also introduced by Einstein's theory. This indicates that the passage of time can be slowed or sped up, depending on the velocities of the two observers.

The "twin paradox," a well-known thought experiment in which two witnesses are used to show time dilation, is one such example. Just picture a pair of identical twins, one of whom stays on Earth while the other travels across space at nearly the speed of light. The twin who goes on the trip will mature more slowly than their counterpart who stays on Earth. Their relative motion causes a time dilation effect, which explains this occurrence.

The discovery of time dilation has profound consequences for how we interpret the cosmos. It casts doubt on Newtonian absoluteness and uniformity in time. Instead, it shows that time is not a static concept but rather a relative one that changes with the movement of observers and their reference frames.

The effects of time slowing down have been observed and tested experimentally, proving their reality. For instance, high-precision atomic clocks on board satellites in Earth's orbit have been shown to run marginally quicker than clocks on Earth's surface because of the satellites' exposure to a weaker gravitational field. Time dilation is also confirmed by measurements of cosmic rays and high-energy particles traveling near to the speed of light.

In addition to theoretical concerns, time dilation has real-world consequences for tools like GPS. GPS satellites experience gravitational time dilation and the effects of their high orbital velocities, which can introduce some timing errors into GPS readings. The satellite clocks are fine-tuned to account for these influences and provide precise positional data.

Time dilation is not only important from a scientific standpoint, but it has also inspired many works of fiction and the entertainment industry. It has become an integral part of literature that speculates on the feasibility of space travel, time travel, and galactic exploration.

The ability of human ingenuity, creativity, and scientific research to uncover time dilation is remarkable. Our prior assumptions about time's linearity and universality have been called into question, leading to a more nuanced understanding of these concepts. This idea has shaped our cosmic exploration and continues to motivate scientists in physics, cosmology, and astrophysics. The more we learn about time dilation, the more we learn about the cosmos and our place in it.

EINSTEIN'S THEORY OF RELATIVITY

When it comes to our knowledge of space, time, and gravity, Einstein's theory of relativity is among the most innovative scientific theories in history. The two main parts of this theory, both of which were proposed by the physicist Albert Einstein in the early 20th century, are the special theory of relativity and the general theory of relativity.

In 1905, Albert Einstein presented his special theory of relativity, which fundamentally altered our conceptualization of the cosmos. It introduced the idea that the laws of physics are the identical for all observers moving at the same speed relative to one another. Objects' behavior at high velocities can now be understood within a framework that challenges the conventional concept of absolute time and space, according to this principle.

In the special theory of relativity, the "relativity of simultaneity" is a crucial principle. This theory proposes that the relative velocity of observers can affect their perception of events occurring at the same time. Events that look simultaneous to one observer may not appear simultaneous to another, for as when two spacecraft pass each other at a substantial fraction of the speed of light. This relativistic effect results from the fact that light travels at a finite speed and that time and distance are relative to the observer.

Time dilation, the idea that time can expand or shrink based on the motion of two observers, was also presented by the theory. When compared to a stationary observer, time passes more slowly for moving objects due to a phenomenon known as time dilation. This effect is amplified as the speed of travel approaches the speed of light. If an astronaut were to spend a given amount of time aboard a spacecraft traveling at nearly the speed of light, upon their return they would have aged less than their identical twin who stayed on Earth. Using extremely precise atomic clocks, this phenomenon has been validated experimentally.

In 1915, Einstein developed the general theory of relativity, which built on the special theory and added the idea of gravity as a curvature of spacetime. Massive objects, such as planets and stars, are thought to cause a curvature in spacetime, forcing nearby things to move along curved pathways. What we experience as gravity is actually just this bending.

Gravitational lensing is a well-known example of a phenomenon that conforms to general relativity's predictions. When light from a distant object, such a galaxy or quasar, travels close to a huge object, like a galaxy cluster or a black hole, it undergoes gravitational lensing. The strong gravitational field of the enormous body causes light to deviate from a straight line. This causes an observer on Earth to see a distortion or a magnification of the faraway item. In 1919, during a solar eclipse, Sir Arthur Eddington's expedition team discovered and confirmed this phenomenon by measuring the deflection of starlight passing near the Sun.

The general theory of relativity also predicted the presence of gravitational waves, which are ripples in spacetime created by the acceleration of enormous objects, in addition to gravitational lensing. Direct evidence for the presence of gravitational waves was provided in 2015 by the Laser Interferometer Gravitational-

Wave Observatory (LIGO), further validating the predictions of general relativity.

Even outside physics, Einstein's theory of relativity has had far-reaching effects in fields like astronomy, cosmology, and technology. Successful uses include satellite navigation systems and the accurate synchronization of atomic clocks, and it has produced a clearer understanding of the nature of space, time, and gravity.

In conclusion, the fundamental ideas of space, time, and gravity were completely rethought thanks to Einstein's theory of relativity. Classical notions of absolute time and space were called into question by the relativity of simultaneity and time dilation introduced by the special theory of relativity. These concepts were developed further and the idea of gravity as a curvature of spacetime was incorporated in the general theory of relativity. Gravitational lensing and gravitational waves, among other phenomena, were both predicted by the theory and later confirmed by experiment. There are many theoretical and practical consequences of Einstein's theory of relativity that continue to impact our view of the universe.

UNDERSTANDING TIME AS A DIMENSION

Understanding time as a dimension goes beyond perceiving it as a linear progression of moments. It involves viewing time as an integral component of the fabric of the universe, similar to the three spatial dimensions of length, width, and height. This perspective arises from the field of theoretical physics, where the concept of time as a dimension is explored in depth, particularly in the context of spacetime.

In classical physics, time is often treated as a separate entity, distinct from the three spatial dimensions. However, in Albert Einstein's theory of relativity, time and space are unified into a single entity known as spacetime. According to this theory, the three spatial dimensions are interconnected with time, forming a four-dimensional continuum where events occur. This understanding revolutionized our perception of time and provided a framework for comprehending its relationship with space.

In the realm of spacetime, time is no longer considered an independent variable but rather a dimension through which objects and events move. Similar to how an object can move through the spatial dimensions, it can also move through the dimension of time. This perspective allows us to visualize the universe as a four-dimensional structure, where objects exist at specific points in both space and time.

To illustrate this concept, let's consider a case study involving a hypothetical spacecraft journey. Imagine a spacecraft traveling from Earth to a distant star system. From the perspective of classical physics, the spacecraft would be viewed as moving through three-dimensional space, with time being a separate parameter to measure the duration of the journey. However, when we embrace the concept of time as a dimension, the spacecraft's trajectory is represented as a path through four-dimensional spacetime.

In this case, the spacecraft's motion is not solely described by its coordinates in three-dimensional space but also by its position along the dimension of time. The spacecraft's journey can be visualized as a curve in four-dimensional spacetime, showing its path through both space and time. This visualization emphasizes the interconnectedness of space and time, highlighting that an object's motion is a result of traversing through both spatial and temporal dimensions.

This understanding of time as a dimension has profound implications for various phenomena, including time dilation and the concept of "warping" spacetime. Time dilation occurs when the passage of time varies for different observers depending on their relative motion or proximity to massive objects. It suggests that time can elapse differently for objects moving at different speeds or experiencing different gravitational fields. This phenomenon has been experimentally verified and is a fundamental aspect of Einstein's theory of relativity.

Another consequence of viewing time as a dimension is the concept of "warping" or bending spacetime. Massive objects, such as planets or black holes, can cause a distortion in the fabric of spacetime. This distortion affects both the spatial and temporal dimensions, altering the motion of objects within that region. For example, near a massive object, time appears to pass more slowly compared to regions further away. This gravitational time dilation is a result of the warping of spacetime by the object's gravitational field.

One notable example that showcases the interplay of time and gravity is the phenomenon of gravitational time dilation observed around black holes. Black holes are extremely massive objects that exert a powerful gravitational pull. Near the event horizon, the boundary beyond which nothing can escape the black hole's gravitational pull, time dilation becomes significant. An observer close to the event horizon would experience time passing much slower compared to an observer further away from the black hole.

To summarize, understanding time as a dimension involves perceiving it as an integral component of spacetime. This perspective unifies time with the three spatial dimensions, allowing us to view the universe as a four-dimensional continuum. Time becomes a dimension through which objects and events move, and our understanding of motion expands to include traversal through both space and time. This conceptual framework has significant implications for phenomena such as time dilation and the warping of spacetime by massive objects. Through the lens of this understanding, we can explore the fascinating interplay between time, space, and gravity, shedding new light on the nature of our universe.

THE IMPORTANCE OF TIME DILATION IN PHYSICS

Einstein's theory of relativity introduces an intriguing new idea in physics: time dilation. It's the apparent slowing or quickening of the passage of time as experienced by observers in various gravitational environments or while in motion. The way we conceive of time, space, and the cosmos have all been altered by this idea. This essay will explore the theoretical underpinnings and practical ramifications of time dilation and its significance in the field of physics. We'll also look closely at a real-world example to show how it works.

Time dilation is based on Einstein's special theory of relativity, which claims that all observers moving at constant velocity experience the same physical laws. This theory postulated that the speed of light in a vacuum is a fixed value that does not change no matter how fast or slow the light's origin or observer is traveling. As a result, the idea of spacetime was born, in which time and space are fused to create a fourth dimension.

Time slows down or speeds up depending on the relative velocities involved in special relativity. Time is said to slow down for moving objects in comparison to a stationary observer under this hypothesis. As speed approaches the speed of light, the effect becomes more dramatic. From special relativity, we get the time dilation formula, which describes the connection between speed and time dilation in terms of the speed of light.

Time dilation has important ramifications for several branches of science, such as astronomy, particle physics, and global positioning systems (GPS).

In astronomy, time dilation has had a profound effect on our perception of the cosmos, making it a unique application. The redshift of light from distant galaxies can be understood by this theory. Due to the expansion of the universe, time slows down for light as it travels across enormous intergalactic distances. Light is redshifted because its wavelength is lengthened by this dilation. Using the redshift, astronomers can determine how far away distant objects are and learn more about the expansion of the universe.

Time dilation is very important in particle accelerators like the Large Hadron Collider (LHC) in particle physics. Particles are propelled to almost the speed of light in these high-energy experiments. Because of temporal dilation, it seems as though their lifespans have lengthened. This enables researchers to monitor and measure particles with extremely short lifetimes that would otherwise vanish too quickly to be observed. Researchers can better comprehend fundamental particles and their interactions by calculating decay rates and particle lifetimes with time dilation taken into account.

When it comes to the Global Positioning System (GPS), which uses accurate timing measurements to identify positions, time dilation is also a crucial factor. The GPS satellites are constantly in motion with respect to ground-based observers. Therefore, time dilation occurs between the satellites' clocks and Earth's clocks. Without taking time dilation into account, GPS coordinates will be off by a large margin. So, the satellite clocks are fine-tuned to maintain precise time synchronization, which in turn permits accurate location measurements.

The Twin Paradox We will use the well-known "Twin Paradox" thought experiment to demonstrate the implications of time dilation. Picture Alice and Bob, two conjoined twins. While Bob travels across space at a considerable fraction of the speed of light, Alice stays on Earth. Time dilation causes Bob's perception of time to slow down as he's traveling. Back on Earth, he discovers that he has aged far more slowly than Alice has.

The twins' relative velocity creates time dilation, creating a paradox because each twin is experiencing events from a distinct reference frame. Due to his great speed, Bob's clock appears to run slower from Alice's point of view. However, Bob thinks Alice's clock back on Earth is ticking slower than it actually is. When Bob returns to Earth, he suddenly shifts reference frames, ending the paradox. Because of this modification, a comparison of clocks shows that Bob has aged less than Alice.

The Twin Paradox illustrates that time is not an absolute quantity but is affected by relative motion, and draws attention to the fundamental nature of time dilation. It demonstrates why high-velocity or gravitational environments necessitate taking time dilation into account.

Einstein's theory of relativity gives rise to a unique physics phenomenon known as time dilation. Our conceptions of how the cosmos works in terms of time and space have been revolutionized as a result. The phenomena has important applications in a number of scientific disciplines, including astronomy, particle physics, and global positioning system technology. Scientists can generate precise forecasts, monitor particle lifetimes, and guarantee correct position determination if they take time dilation into consideration. The Twin Paradox is an intriguing case study that shows how time dilation can impact a hypothetical situation involving a pair of identical twins. Time dilation shows how time and motion are intrinsically linked and undermines our intuitive conception of time as a constant. Time dilation is a vital and fascinating part of our understanding of the cosmos, and we have only just begun to scratch the surface of its implications.

THE POSTULATES OF SPECIAL RELATIVITY

The postulates of special relativity are fundamental principles that form the basis of Albert Einstein's theory of relativity. These postulates provide the framework for understanding the behavior of objects moving at speeds close to the speed of light and have far-reaching implications for our understanding of space, time, and the nature of physical reality.

The first postulate of special relativity, known as the principle of relativity, states that the laws of physics are the same in all inertial reference frames. An inertial reference frame is a coordinate system that is not accelerating but moving at a constant velocity. This postulate implies that there is no privileged or preferred reference frame in the universe, and the laws of physics are consistent across all such frames. It challenges the classical notion of absolute space and time, suggesting that physical phenomena are relative and depend on the observer's perspective.

The second postulate, known as the constancy of the speed of light, states that the speed of light in a vacuum is always constant, regardless of the motion of the source or the observer. This means that the speed of light, denoted by "c," has the same value for all observers, regardless of their relative motion. This postulate has profound implications, as it contradicts the classical notion of Galilean relativity, where the speed of light would be expected to vary depending on the motion of the source or observer.

The constancy of the speed of light forms the foundation for understanding time dilation and length contraction.

To illustrate the concepts of special relativity and the implications of the postulates, let's consider a case study involving two observers, Alice and Bob. Imagine Alice is on a stationary spaceship, and Bob is on Earth. They have synchronized clocks, and Alice sets off in her spaceship traveling at a significant fraction of the speed of light relative to Earth.

According to the principle of relativity, Alice can consider herself at rest and view Earth as moving away from her. Likewise, Bob on Earth can consider himself at rest and view Alice's spaceship as moving away. Both Alice and Bob can claim to be at rest within their respective reference frames.

However, according to the constancy of the speed of light, Alice and Bob will observe different outcomes. Let's say Alice has a powerful flashlight, and she shines a beam of light towards a mirror at the other end of her spaceship. According to Alice, the light beam will travel in a straight line and reach the mirror and bounce back to her.

Now, from Bob's perspective on Earth, he will also see the light beam traveling away from Alice's spaceship towards the mirror. However, Bob will observe that the light beam takes a longer path due to the motion of the spaceship. This is because while the light beam is traveling towards the mirror, the mirror itself is moving away from the point where the light was emitted.

Since the speed of light is constant for both Alice and Bob, the light beam will take the same amount of time to reach the mirror and return to Alice, as observed from both reference frames. This phenomenon is known as time dilation, where time appears to move slower for a moving object relative to a stationary observer.

From Alice's perspective, her clock will tick at a normal rate, and she will perceive time passing as usual. However, from Bob's perspective, he will observe that Alice's clock appears to be running slower than his own clock on Earth. This is due to the relative motion between Alice and Bob and the time dilation effect predicted by special relativity.

The case study highlights the key concepts of special relativity and how they manifest in the observations of different observers in relative motion. The postulates of special relativity challenge our classical understanding of space, time, and motion and provide a framework for understanding the behavior of objects moving at high speeds. The constancy of the speed of light and the principle of relativity have been rigorously tested and confirmed through numerous experiments and observations, solidifying their place in the foundations of modern physics.

TIME DILATION AT HIGH SPEEDS

An intriguing aspect of relativity theory, and of special relativity in particular, is the phenomenon of time dilation at high velocities. This theory states that as a moving object approaches the speed of light, it experiences a slowing of time in comparison to a stationary observer. This phenomenon, which physicists refer to as time dilation, is central to our understanding of the universe.

Time dilation at high velocities requires an understanding of the fundamentals of special relativity. One of the basic tenets of this theory is that the physical laws do not change when moving from one observer to another at a constant speed. This indicates that the observed physical phenomena will be the same regardless of the relative motion of the observer, as there is no absolute reference frame.

The speed of light in a vacuum, indicated by "c," is thought to be the maximum possible velocity under special relativity. According to Einstein's theory, the speed of light remains the same for all observers, no matter how they are moving with respect to one another. Time intervals measured by an object traveling close to the speed of light will be stretched or elongated in comparison to those recorded by an observer standing still.

The relationship between space and time gives rise to the idea of time dilation. The concept of spacetime, in which space and time are inextricably linked, is introduced by special relativity. An object's relative velocity through spacetime modifies the flow of time. When compared to a stationary observer, time passes more slowly for a moving object.

Lorentz factor gives the equation for time dilation:

$\gamma = 1 / \sqrt{(1 - v^2/c^2)}$

The time dilation factor is denoted by, the speed of the moving object is denoted by v, and the speed of light is denoted by c. Significant time dilation occurs when the object's velocity is close to the speed of light because v/c becomes a large fraction and the Lorentz factor approaches infinity.

Time dilation at high speeds can be demonstrated by examining a specific scenario, the so-called "Twin Paradox." We have Alice and Bob, who are identical twins. While Bob stays on Earth, Alice makes plans to travel through space aboard a spaceship. For a large portion of her trip, Alice travels at a speed close to one-ninth the speed of light, such as 0.9c.

Time would seem to pass more slowly for Alice as she travels away from Earth compared to Bob back on the ground, in accordance with special relativity. From Bob's vantage point, Alice's clock moves more slowly than his own back on Earth. If this is the case, then Alice will have aged less than Bob, who stayed on Earth and aged normally, by the time she returns from her trip.

The time slowing effect can be measured with the use of the Lorentz factor equation. If Alice is traveling at a speed of 0.9c, then the Lorentz factor is determined as follows:

$$\gamma = 1 / \sqrt{(1 - (0.9c)^2/c^2)} \approx 2.294$$

In other words, relative to Bob's time on Earth, Alice's is stretched out by a factor of about 2.294. If Alice left on a trip that would take one year by her reckoning, Bob would calculate that she has actually been away from Earth for about 2.294 years. This is evidence of the extreme time slowing down that occurs when an object travels at great speeds.

An intriguing facet of time dilation is the symmetry of perception, which is highlighted in the Twin Paradox case study. If Alice were on the spaceship, she would see Bob's time on Earth slow down from her vantage point. Bob's clock would appear to be ticking more slowly in her mind than her own. When Alice finally gets back to Earth, though, the symmetry is thrown off and Bob ends up being the older of the two.

The effects of acceleration and changes in reference frames allow us to reconcile this seeming contradiction in the Twin Paradox. When comparing the time perceptions of a "traveling twin" with a "stationary twin," acceleration is a key factor. This is a crucial element when trying to understand time dilation in real-world circumstances, but it requires a more in-depth analysis than can be provided here.

Multiple tests and observations have verified the existence of time dilation at high speeds. For example, the Large Hadron Collider (LHC) and other particle accelerators propel particles to speeds near the speed of light. By observing that the lifetimes of the accelerated particles are longer than those of their stationary counterparts, these experiments have revealed empirical proof of time dilation.

In addition, GPS accounts for time dilation caused by the fast speeds at which satellites travel across space. Because GPS satellites are moving with respect to Earth, their onboard clocks have a tiny advantage over terrestrial timepieces. The GPS calculations need to account for time dilation to get reliable position data.

Finally, time dilation at high speeds is an extremely significant implication of Einstein's special theory of relativity. Relative to a stationary observer, time appears to slow down as an object approaches the speed of light. Experiments and observations have both corroborated this phenomena, lending credence to the theory. The Twin Paradox is an excellent example of time dilation because it shows how the aging process might differ for moving and stationary observers. The effects of time dilation on the passage of time, the structure of spacetime, and the behavior of moving objects have profound implications for our knowledge of these topics.

LORENTZ TRANSFORMATIONS

Lorentz transformations, conceived of by Dutch physicist Hendrik Lorentz in the late 19th and early 20th centuries, are central to the special relativity theory. For observers traveling at high relative velocities, these transformations explain how space and time appear to them. They are the mathematical underpinning that allows the constancy of the speed of light in all inertial reference frames to be consistent with the laws of relativity.

The two postulates of special relativity are prerequisites for comprehending Lorentz transformations. The first postulate states that all inertial reference frames, or reference frames that move at a constant velocity with respect to one another, have the identical set of physical laws. According to the second postulate, the relative velocity of a source and an observer in a vacuum remains constant.

One implication of these postulates is that the relative measurement of time and distance depends on the frame of reference used by the observer. Mathematical equations provided by Lorentz transformations link observations of space and time from several reference frames. They shed light on the discrepancies between these readings made by observers traveling at various speeds.

Multiple fundamental ideas are intertwined in the Lorentz

transformations, such as the relativity of simultaneity, length contraction, and time dilation. Time dilation describes the apparent lengthening of time periods for moving observers relative to those who remain still. This phenomenon results from the fixed speed of light and the motion of the observers. The Lorentz transformations state that time dilation increases as an object's velocity increases toward the speed of light.

When a stationary observer looks at a moving item, the object seems shorter in the direction of motion, a phenomenon known as "length contraction." Constant light speed and relative motion between observers also contribute to this impression. According to the Lorentz transformations, the length contraction of an object grows increasingly noticeable as its speed approaches the speed of light.

Another important facet of Lorentz transformations is the relativity of simultaneity. It says that simultaneity is relative, and that events that look simultaneous in one reference frame may not appear simultaneous in another. The length of time it takes light to travel from an event to an observer creates this phenomena. Thus, the idea of simultaneous events becomes contextual and dependent on the frame of reference of the observer.

Let's take the twin dilemma as an illustration of how Lorentz transformations can be used in practice. The twin paradox is a hypothetical situation in which one identical twin leaves Earth and speeds into space while the other twin stays here. Special relativity states that one of the twins will age more slowly than the other because of time dilation.

Let's pretend the twin who is able to travel the fastest sets off on a mission to a nearby star system. From their vantage point, it might take no more than a few years to make the journey. When they go back to Earth, though, they discover that a lot of time has passed and that their twin has aged noticeably more than they have. Lorentz transformations are responsible for the time dilation phenomenon, which causes this disparity in aging.

Lorentz transformations accurately explain the connection between time, velocity, and the relative motion of observers, as demonstrated by the twin paradox. It exemplifies the far-reaching effects of special relativity on how we think about time and space and how useful the theory can be in real-world situations.

In conclusion, Lorentz transformations are a set of mathematical equations that describe the effect of changing observers' reference frames on spacetime measurements. They lay the groundwork for comprehending concepts central to special relativity, such as time slowing down, space shrinking, and simultaneity being relative. Lorentz transformations have important applications in many fields, from space travel and particle physics to high-velocity events, and are crucial in reconciling relativity's principles with the constancy of the speed of light. Scientists and researchers can gain a better grasp of the nature of reality by applying Lorentz transformations to precisely describe and anticipate the behavior of physical systems in different reference frames.

TIME DILATION IN EVERYDAY LIFE

Einstein's theory of relativity introduces the concept of time dilation, which is typically linked with extremely rare circumstances such as traveling at near-light speeds or in the proximity of huge objects. Time dilation, however, has implications outside of the domains of astronomy and the laboratory. The effects of time dilation on our daily lives can be subtle, but they are real all the same. The interplay among gravity, velocity, and the passage of time gives rise to this phenomena.

Let's look at the theory behind time dilation to see how it works in practice. The theory of relativity states that observers in various motions or gravitational fields will perceive time differently. There are two primary kinds of time slowing, and they have to do with speed.

Time slows down due to gravitational potential variations. As the strength of a gravitational field increases, the passage of time slows down. A clock at the foot of a mountain, for instance, will tick marginally faster than one at the peak. Time slows down at greater heights because the gravitational force is weaker there. Time dilation is also amplified near extremely big objects like black holes because of their exceptionally powerful gravitational fields.

Moving at a high speed in relation to another item causes what is called velocity time dilation or time dilation owing to relative motion. When an object is moving at close to the speed of light, time slows down for it relative to an observer in a stationary reference frame, according to the theory of relativity. This implies that time seems to move more slowly for a fast-moving observer than for a stationary one. However, at usual speeds, this impact is insignificant, and significant time variations appear only at velocities near the speed of light.

Time dilation has measurable effects and real-world relevance, notwithstanding their relative insignificance in most situations. The Global Positioning System (GPS) is a prominent case in point. To pinpoint their location on Earth, GPS receivers rely on extremely accurate timing signals. But time dilation plays a role because satellites orbiting Earth at fast speeds. When viewed from Earth, the satellites travel at a speed of about 14,000 kilometers per hour. Therefore, their watches undergo velocity time dilation, which, if ignored, would cause GPS positioning inaccuracies of many kilometers each day. To prevent inaccurate positioning data, scientists and engineers must modify the satellite clocks to account for time dilation.

The "Twin Paradox" is an intriguing case study that exemplifies the consequences of time dilation in real life. In this hypothetical scenario, one set of twins stays on Earth while the other travels through space at breakneck speed. The twin who has been away from Earth for a long time will age more slowly than their sibling who has stayed put. The time dilation the traveling twin experiences as a result of their rapid mobility is the root cause of this apparent paradox.

Consider the hypothetical situation of Alex and Sam, who are identical twins. While Sam travels across space at a considerable fraction of the speed of light, Alex stays on Earth. Sam takes a trip to a nearby star system and comes back to Earth thereafter. Due to time dilation brought on by Sam's high velocity relative to Earth, their clock runs slower than Alex's on Earth. When Sam finally returns to Earth after several years, he and his family discover that they have aged far more slowly than Alex has. The time dilation effect from relative motion and the intriguing effects on aging are demonstrated here.

These are only a few of the many scientific and technological sectors that could be affected by time dilation. For instance, particle accelerators use time dilation to rapidly accelerate subatomic particles. The longer particles spend in the accelerator, the more time dilation they encounter, giving researchers a more in-depth look at their behavior.

In addition, knowing about time dilation helps us make sense of cosmological events and other astrophysical phenomena. It's a major factor in things like gravitational waves and how things act when they're close to extremely large stars or planets.

The existence and impact of time dilation have been experimentally established, despite the fact that it is not readily perceptible in everyday life without precise measurements. In extreme conditions or when traveling close to the speed of light, the effects become more pronounced. Time dilation, however, poses a serious challenge to our natural tendency to think of time as a fixed and unchanging entity.

Einstein's theory of relativity introduces the concept of time dilation, which has implications well beyond the realm of astronomy. Time can be subtly altered in different situations due to factors like gravity time dilation and velocity time dilation. As an example of the need of taking time dilation into account, consider the adjustment of satellite clocks in GPS systems. The fascinating Twin Paradox exemplifies the far-reaching effects of time dilation on biological aging and relative velocity. The study of time dilation improves our understanding of the cosmos, aids in the development of new technologies, and questions our assumptions about the nature of time.

THE TWIN PARADOX

Einstein's theory of relativity predicts the existence of a phenomena known as time dilation, which can be investigated through the use of a thought experiment known as the Twin Paradox. The twin who stays on Earth is left behind as his or her sibling speeds toward space. When the two are reunited, they discover that the twin who left Earth has aged more slowly than the twin who stayed. This paradox makes us question our fundamental assumptions about how time and the universe work.

Studying special relativity and the idea of time dilation is essential for solving the Twin Paradox. Special relativity states that time's passage is relative and subject to change based on variables like speed and gravity. Time slows down when one item or observer travels relative to another at a high percentage of the speed of light. For a moving item, time passes more slowly than it does for a stationary one as its speed increases.

Take Alice and Bob, for example, who are identical twins. They settle on an experiment where Bob stays on Earth and Alice goes on a hyperspace mission. Alice will visit a nearby star system and then come back to Earth, as per their agreement. Time will slow down for Alice in comparison to Bob on Earth as she accelerates and achieves high velocities, as predicted by special relativity.

Upon her return to Earth, Alice discovers that she has aged more slowly than Bob. Time dilation provides an explanation for this apparently contradictory result. Time seems to go more slowly for Alice when she travels at high speeds. Her body clock, in other words, is operating at a slower rate than Bob's on Earth. Therefore, less time will have elapsed for her when she returns, in comparison to Bob.

Let's use an example to further clarify this point. Let's pretend that Alice departs Earth when she and Bob are both 30 years old. Almost at the speed of light, she reaches a star system only 10 light-years distant. Due to time dilation, the trip appears to take only a few years from her vantage point. If it takes her three years, so be it.

Bob, meantime, is back on Earth and has seen 13 years pass. This is because, from Bob's vantage point, Alice has gone 20 light-years away from Earth and back, twice. As a result of time dilation, Alice clocked in at three years whereas Bob clocked in at thirteen, as predicted by the theory of special relativity.

Alice will be 33 years old and Bob will be 43 when she finally gets back to Earth. Alice's high-velocity voyage caused time dilation, which accounts for the 10-year age gap. The interesting nature of the Twin Paradox is demonstrated here by the fact that two people born at the same time can age at different rates due to time dilation.

So, how do we explain this contradiction? The key to solving the problem is realizing that it is not a zero-sum game. Alice speeds up to a high speed and then slows down to return to Earth, while Bob's speed on the ground remains constant. According to general relativity theory, speeding up or slowing down can alter how quickly or slowly time goes by. When Alice alters her speed, her reference frame shifts, and her perception of time diverges from Bob's.

There are more aspects to think about, such as gravity's influence. Time dilation calculations can get even more involved if the trip takes the traveler through locations with varying gravitational fields. Understanding the inner workings of the Twin Paradox requires taking into consideration the fact that gravity can affect the flow of time.

The Twin Paradox is a thought experiment, not a real-world circumstance amenable to actual testing. However, atomic clock and high-velocity particle studies have supplied data to back up the theory of time dilation. These results show that time dilation is a genuine phenomenon that occurs in accordance with relativity theory.

The Twin Paradox calls into question our natural assumption that time is unchanging and draws attention to the complex interplay between these three variables. It challenges us to examine our assumptions about the world and the ramifications of Einstein's relativity theories. The paradox may be difficult to understand at first, but it finally reveals the incredible scope of the universe and the interesting ways in which motion and gravity may distort time.

The Twin Paradox is a thought experiment that examines the concept of time dilation in the setting of two identical twins, one of whom is moving at high velocity while the other is standing still. The paradox results from the fact that the twin who is constantly on the move ages more slowly than the twin who stays put. One must be familiar with time dilation and the effects of velocity and acceleration in order to make sense of the Twin Paradox. The paradox may test our common sense about how time works, but it also reveals how intriguing the universe is and how time, motion, and gravity all interact in complex ways.

THE EQUIVALENCE PRINCIPLE

A strong connection between gravity and acceleration is implied by the equivalence principle, a basic notion in physics. Albert Einstein first proposed it as part of his theory of general relativity, and it has far-reaching consequences for how we conceptualize space, time, and gravity. The concept, which forms the basis for Einstein's theory of gravity, argues that the effects of gravity and acceleration are indistinguishable.

Understanding gravity and its impact on matter and the fabric of spacetime is crucial to getting a handle on the equivalence principle. The force responsible for the attraction of all massive objects to each other is called gravity. It's the force that causes everyday occurrences like dropped objects to fall to the earth and the movement of heavenly bodies across the cosmos. The attraction between two massive objects is proportional to the product of their masses and inversely proportional to the square of the distance between them, as described by Isaac Newton's theory of gravity.

Revolutionizing our knowledge of gravity, Einstein's general theory of relativity proposed that mass and energy induce a curvature in spacetime that we now refer to as gravity. This concept postulates that huge objects, such as planets and stars, generate a gravitational field that warps spacetime in their vicinity. As they move through this warped universe, other objects feel a pull that we call gravity.

Einstein's observation that the consequences of gravity and acceleration are quantitatively equivalent gave rise to the equivalence principle. An observer in an enclosed elevator is used to illustrate this principle. If the elevator is on the ground level, the spectator will feel the pull of gravity as they descend. When an elevator travels upward at a steady rate in deep space, an observer within the elevator will feel a force pressing them to the floor, similar to the force of gravity, even though there is no gravitational effect.

The equivalence principle states that an elevator passenger is not capable of telling the difference between the two events. From the perspective of the person in the elevator, the effects of gravity and acceleration are the same, and they cannot conduct any experiment or make any measurement to determine whether they are in a gravitational field or experiencing acceleration.

This principle has far-reaching consequences for how we conceptualize gravity and spacetime. It implies that an object's sense of gravity is not the result of a direct force acting upon it, but rather a consequence of the object's motion through curved spacetime. That is to say, gravity is not a force that acts between things, but rather the bending of spacetime.

High-quality experimental confirmation of the equivalence principle is available. The Eötvös experiment, undertaken by physicist Robert Dicke in 1959, is a well-known example of this type of experiment. The accelerations of two test masses of different compositions were compared while they were in the presence of the Earth's gravitational field. If gravity were a force operating directly on the masses, then the ratio of their accelerations would change with their composition since the gravitational force experienced by the masses would depend on their composition. Within the experimental tolerances, however, the ratio was shown to have remained unchanged, providing strong proof for the equivalence principle.

The phenomenon of gravitational time dilation also serves to explain the equivalence principle. Clocks in a gravitational field are predicted to run slower than those in areas with less gravitational pull, per general relativity. The renowned Hafele-Keating experiment from 1971 is just one of many that have verified this effect. In this study, atomic clocks were transported through planes flying in opposite directions across the globe. When the planes returned, their clocks showed discernible time changes because of the varied gravitational forces they had experienced at different altitudes. This time slowing effect is in line with general relativity's predictions and lends credence to the equality principle.

The equivalence principle is fundamental to general relativity and has far-reaching consequences for our worldview. Einstein's theory of gravity, in which gravity is not a force acting between objects but rather the curvature of spacetime, is based on this idea. Black holes, gravitational waves, and the expansion of the universe may all be traced back to this theory, which has paved the way for many theoretical and practical advances in physics. The equivalence principle has altered our understanding of spacetime and the physical laws that govern the universe by highlighting the essential link between gravity and acceleration.

CURVATURE OF SPACETIME

Curvature of spacetime is central to Einstein's general theory of relativity, which fundamentally altered our perception of gravity and the cosmos. This theory proposes that gravity results from the curvature of spacetime caused by the presence of mass and energy in the universe. According to this theory, huge things not only draw in and reroute other objects, but also distort the very fabric of spacetime itself.

Understanding the connection between space and time within the context of general relativity is crucial to grasping the idea of the curvature of spacetime. According to this concept, time is the fourth dimension of a four-dimensional spacetime continuum that also includes the traditional three dimensions of space (length, width, and height). General relativity merges space and time into a single entity, highlighting their interconnectedness, rather than perceiving them as separate entities.

Gravitational field is what we call the warping of spacetime caused by the presence of mass and energy. According to general relativity, spacetime is curved in response to the existence of matter, and this in turn directs the motion of matter. The theory's explanation of gravity rests on this dynamic between matter and spacetime.

Consider a big object, like a star, in space as an example of the curvature of spacetime. The star's mass warps spacetime around it, creating a gravitational field, in accordance with general relativity. As a result, anything near the star will be affected by the curvature, and it will cause it to move along a curved path or orbit.

Imagine a planet in orbit around a star, and you can see it moving in a perfect circle because of the star's gravitational pull. The mass of the star causes spacetime to curve, therefore this motion is not truly circular under general relativity but rather an orbit. The planet is "rolling" along the warped spacetime produced by the star.

The planet's curved path results directly from spacetime's curvature. A "dip" or depression in spacetime is produced by the star's presence due to its mass and energy. The planet's path along its orbit follows the arc of this "dip," causing it to spiral inward toward the star even as it continues to travel forward. A stable orbit around the star is the outcome of the combined effects of gravity and the spacetime curve.

In addition, spacetime curvature is not confined to the regions immediately surrounding big objects. It permeates everything, even the speed of light, throughout the cosmos. The shape of spacetime has an effect on even massless photons. Light's path is curved by the black hole's enormous gravitational attraction when it travels close to such a large object.

Another spectacular manifestation of the curvature of spacetime, gravitational lensing occurs when light is deflected by the gravity of enormous objects. Gravitational lensing observations have helped us better comprehend gravity and the structure of the cosmos by providing empirical evidence in favor of the predictions of general relativity.

In conclusion, Einstein's theory of general relativity relies heavily on the concept of spacetime curvature. It explains how gravity arises from the presence of mass and energy, which generates a curvature in spacetime. The orbits of planets around stars and the deflection of light around huge objects are both governed by the curvature of spacetime. General relativity provides a unifying framework for studying gravity and the cosmological and microscopic behavior of the universe by taking into account the interaction between matter and spacetime.

GRAVITATIONAL TIME DILATION

The effect of gravity on the passage of time, known as gravitational time dilation, is a fascinating phenomenon in physics. Einstein's general theory of relativity states that any sufficiently massive object, such as a planet or a black hole, can bend space and time around it. That's why an observer's perception of the passage of time might vary with the magnitude of the gravitational field they're subjected to. Numerous tests and observations have validated this idea, which has far-reaching ramifications for our cosmological understanding.

Understanding general relativity is crucial for grasping the idea of gravitational time dilation. This theory proposes that mass and energy induce a curvature in the fabric of spacetime, rather than gravity being a force acting between things. There is a "dimple" in spacetime created by massive objects like stars and planets, which forces nearby objects to move in ellipses. The passage of time is likewise altered by this curvature, along with the speed of objects.

Gravity affects both space and time, which results in the phenomenon of gravitational time dilation. Time moves more slowly in locations with a high gravitational field, such as those near huge astronomical planets. This means that clocks in regions of higher gravitational fields will run more slowly than those in regions of lower fields.

Gravitational time dilation can be understood by contrasting the rate of time passing on Earth's surface with that of a satellite in orbit. High-precision atomic clocks are used to keep satellites, including those used for GPS systems, in time with terrestrial clocks. These clocks must be exceedingly precise so that GPS signals may be relied upon.

The gravitational field is reduced for the satellite because of its greater distance from the Earth's surface. Since the satellite is further away from the Earth, the gravitational field is weaker, hence the satellite's clock will run slightly faster than a surface clock. Although seemingly insignificant, this time discrepancy adds up over time and can cause GPS readings to be inaccurate.

The discrepancy in gravitational time dilation must be accounted for by scientists in order to correct for this anomaly. Timepieces in space are set to lag behind those on Earth by a fraction of a second. This tweak keeps the satellite clocks and ground clocks in sync, which is necessary for GPS to provide precise location data.

The passage of time around a black hole is another well-known instance of gravitational time dilation. Nothing, not even light, can escape the black hole's tremendous gravitational attraction because black holes are so immensely dense. As one gets closer to a black hole's event horizon, time slows down considerably. If you were near the black hole's event horizon, time would seem to move far more slowly to an outside observer than it would to you.

The "twin paradox," in which one identical pair of twins visits a black hole on a spaceship while the other stays on Earth, is an example of this effect. After what seems like a short trip, the twin who stayed on Earth was shocked to see how much older their sibling had gotten while they were away. This occurs because of time dilation brought on by the black hole's powerful gravitational pull.

Multiple tests and observations have established the existence of gravitational time dilation. In order to observe the effects of gravitational time dilation, atomic clocks have been installed aboard airplanes and flown at high altitudes. The results have always demonstrated that airplane clocks are slightly faster than ground-based clocks. This gives hard proof for the existence of gravitational time dilation, confirming the predictions of general relativity.

In addition, the consequences of gravitational time dilation for our cosmological understanding are substantial. This finding is consistent with the idea that gravitational fields can alter the passage of time. This idea has been fundamental to our grasp of cosmology, particularly in regards to the study of time's behavior in the vicinity of extremely enormous objects like black holes.

Time dilation due to gravity is an amazing prediction of Einstein's general theory of relativity. Time flows at varying rates in places with different gravitational fields, which results from the fact that gravity affects both space and time. Experiments and observations, such as those with atomic clocks on satellites and airplanes, have verified the existence of this phenomenon. As well as contributing to our understanding of the cosmos and time itself, the phenomenon of gravitational time dilation has significant practical implications, such as ensuring the proper operation of GPS systems.

BLACK HOLES AND TIME DILATION

Both scientists and laypeople are fascinated by black holes, a mysterious cosmic phenomenon predicted by Albert Einstein's theory of general relativity. Their link to time dilation is just one of many fascinating aspects of these objects. Black holes produce extreme conditions where the effects of gravity and high velocities on the flow of time become extremely obvious, giving rise to the phenomena of time dilation.

It is essential to go into the fundamental concepts of general relativity to grasp the concept of time dilation in the context of black holes. This hypothesis proposes that gravity is a result of the presence of mass and energy, and not only a force acting between heavy objects. When a huge object exerts enough gravitational force, it warps spacetime around it.

Nothing, not even light, can escape the pull of a black hole because the gravity there is so strong. They are produced from the mass-concentrated leftovers of huge stars that have collapsed under their own gravity. Time dilation is just one of the extraordinary results of being close to a black hole, where the gravitational field is extremely intense.

Near a black hole, time slows down because of the extreme curvature of spacetime caused by its enormous mass. As an observer gets closer to the black hole's event horizon (the point of no return from which nothing can escape), the black hole's gravitational pull strengthens. Therefore, local clocks run slower than those of an observer located further away or in a weaker gravitational field. The passage of time slows down due to the pull of gravity.

Time slows down in the vicinity of black holes, and it's easier to understand the notion with an example. Picture Alice and Bob, two observers, on opposite sides of the universe from a black hole. While Alice is quite some distance from the black hole's event horizon, Bob is rather near it. Each one of their clocks is incredibly accurate and keeps perfect time.

Because the gravitational pull on Alice is less intense, she ages normally. Time passes properly for her, and she hears the clock ticking in the same rhythm every day. But time is different for Bob, who is close to the black hole and hence subject to its powerful gravitational field.

Bob's clock begins to slow down in relation to Alice's as he approaches the black hole. Bob experiences time dilation since the gravitational field is significantly stronger for him close to the black hole. Bob's perception of time passes more slowly than Alice's as a result of the extremely warped spacetime produced by the strong gravity.

Time dilation would be considerably more pronounced if Bob went even closer to the black hole. Bob's time would appear to stop from Alice's point of view as he eventually approaches the event horizon. The enormous gravitational attraction near the event horizon causes time to stop for Bob, causing him to freeze in place.

However, Bob's perspective on time has not changed. Time keeps on ticking at the same rate for him. The relativity of time dilation is illustrated by this phenomenon. The magnitude of time dilation that an observer feels depends on the strength of the gravitational field they are in.

Many independent lines of evidence support the idea that time slows down close to black holes. One such impact is gravitational redshift, where light from objects near a black hole is distorted to longer wavelengths due to the time dilation effect. The effect of gravity on time is made clear by this redshift.

The effects of time dilation near black holes on our view of the cosmos are far-reaching. It hints that time is malleable and subject to the pull of gravity, rather than being a rigid, unchanging entity. Space and time are inextricably woven together, and this fact sheds light on the close relationship between the two.

There are exciting issues to be asked regarding the nature of time and its behavior in the face of strong gravitational forces, which are heightened by the dramatic time dilation around black holes. It paves the way for us to probe the limits of our knowledge, which could yield fresh understanding of the cosmos' fundamental nature.

In conclusion, time dilation and black holes are closely related phenomena. Observers close to black holes experience gravitational time dilation due to the region's high gravitational fields. The relative character of time is demonstrated by this phenomenon, which has been confirmed experimentally and through observations. To further understand how time dilation works and how the strength of the gravitational field affects it, consider the case of an observer near a black hole. Our knowledge of gravity, spacetime, and the very essence of time is enriched by research into time dilation around black holes.

TESTING GENERAL RELATIVITY

Validating and improving Albert Einstein's theory of general relativity through testing is an important scientific undertaking. By arguing that gravity arises from the curvature of spacetime brought about by mass and energy, general relativity radically altered our conceptions of gravity, space, and time. Over the years since general relativity was first proposed, its predictions have been subjected to a plethora of testing, both in the lab and through careful observation. Many predictions of the theory have been verified by these experiments, and they have also opened up new avenues for extending our knowledge of the cosmos.

Light bending in the presence of a big object is a notable prediction of general relativity. Gravitational lensing describes what happens when the course of light is bent by a huge astronomical body. As a result of multiple astronomical investigations, gravitational lensing has been detected and confirmed, lending support to the veracity of general relativity.

The Einstein Cross is a famous example of gravitational lensing. This is an example of gravitational lensing, in which a foreground galaxy distorts the image of a distant quasar billions of light-years away. Multiple images of the same quasar arise due to the galaxy's gravitational pull bending the path of light from the quasar. This finding provides a strong validation of general relativity and Einstein's prediction that gravity can alter the path of light.

General relativity also makes the crucial prediction that time slows down in gravitational fields. Stronger gravitational forces, the idea suggests, cause time to pass more slowly there. Multiple tests and observations, including the well-known Hafele-Keating experiment, have corroborated this conclusion. Atomic clocks were carried on commercial aircraft flying in opposite directions around the world for this experiment. Clocks on board the aircraft, which were exposed to weaker gravitational fields due to their height, were found to run fractionally faster than those stored on the ground. Time dilation, as predicted by general relativity, was supported by this finding.

In addition, testing general relativity can be done by the observation of gravitational waves, which are distortions in the fabric of spacetime brought on by the acceleration of enormous objects. When two black holes merged in 2015, they released gravitational waves, which were first detected by the Laser Interferometer Gravitational-Wave Observatory (LIGO). This finding not only offered hard evidence for the veracity of general relativity's predictions regarding the emission and propagation of gravitational waves, but it also established the existence of gravitational waves.

Measurements of planetary orbit precession, the timing of satellites in orbit, and the redshift of light due to gravity are all additional tests of general relativity. General relativity is one of the most successful and accurate theories in physics, and its predictions have been repeatedly verified by experiments and observations.

Despite its robustness in the face of empirical scrutiny, general relativity is not the final word on the subject of gravity. A theory of quantum gravity, which would unify general relativity with quantum mechanics, is still a topic of ongoing investigation.

Overall, verifying and improving our knowledge of the cosmos depends on our ability to put general relativity to the test. Many of general relativity's predictions, such as gravitational lensing, time dilation, and the presence of gravitational waves, have been verified by experiments, observations, and theoretical analysis. These experiments have not only proven the theory correct, but also expanded our understanding of the complex interplay between gravity, spacetime, and the cosmos. More and better testing of general relativity will surely help us learn more about the cosmos and its inner workings as science and technology continue to improve.

EFFECTS OF GRAVITY ON TIME

Gravity, being one of the universe's fundamental forces, has a tremendous effect on the fabric of space-time, including the flow of time itself. Gravity, according to Einstein's general theory of relativity, is a curvature of space-time created by heavy objects rather than a force that pulls objects towards each other. This curvature affects both matter and light, resulting in intriguing consequences on the passage of time.

The curvature of space-time becomes substantial in the presence of a big object, such as a planet, star, or black hole. This curvature impacts object motion and the perception of time. The higher the object's mass and density, the stronger its gravitational field and the greater the effect on time. This is referred to as gravitational time dilation.

Gravitational time dilation happens as a result of space-time warping near big objects. When an object moves in a gravitational field, its path follows the curvature of space-time, which influences how quickly time passes for that object. Simply put, the closer an object is to a huge body, the slower time passes for it in comparison to an object farther away from the gravitational pull.

Consider a hypothetical scenario involving two observers, Alice and Bob, who are in distinct gravitational conditions. Alice is on Earth, and Bob is aboard a spacecraft orbiting a gigantic black hole. The tremendous gravitational pull of a black hole generates severe time dilation in its surroundings.

Time appears to run normally from Alice's perspective on Earth. However, from Bob's vantage point near the black hole, time appears to slow substantially. The intense gravitational field close to the black hole induces a distortion in space-time, resulting in a slower passage of time for anything inside that zone.

Assume Alice and Bob sync their clocks before Bob begins his trek toward the black hole. Alice notices that Bob's clock appears to tick slower than her own as his spacecraft approaches the black hole. This means that for every second Alice spends, Bob spends less time. As Bob approaches the black hole, the time dilation grows more acute, and the discrepancy in their clocks becomes more obvious.

From Bob's point of view, time moves normally within his spacecraft. When he returns to Earth, he sees that Alice's clock appears to be ticking faster than his own. This finding emphasizes the reciprocal nature of time dilation—the passage of time appears to be slower for Bob as seen by Alice, yet from Bob's perspective, Alice's clock appears to be ticking faster.

The gravitational time dilation found around black holes is scientifically confirmed rather than speculative. The effects of gravity on time, for example, have been established by exact measurements using atomic clocks. Scientists conducted experiments by positioning atomic clocks at various altitudes where the strength of the gravitational field fluctuates slightly. Clocks situated closer to the Earth's surface, which experience a larger gravitational pull, were discovered to run slightly slower than those placed at higher elevations.

This gravitational time dilation phenomena is not restricted to severe conditions such as black holes, but may also be observed on a smaller scale. Even a small variation in gravitational field intensity between the Earth's surface and the top of a mountain, for example, can have a minute time dilation effect. These minor changes show the complex link between gravity and the passage of time.

Gravity's effects on time have ramifications for space exploration and satellite-based technologies. GPS satellites, which orbit the Earth at great altitudes, are subject to lesser gravitational fields than clocks on the Earth's surface. As a result, GPS satellite clocks tick slightly quicker than those on the ground. If the time dilation effect was not taken into consideration, GPS devices would create mistakes, resulting in inaccurate location estimations.

Finally, the effects of gravity on time, referred to as gravitational time dilation, are caused by the curvature of space-time near big objects. The stronger the gravitational field appears to be, the slower time appears to pass. This phenomenon has been empirically validated and seen in a variety of circumstances, ranging from black holes to differences in gravitational field intensity between sites on Earth. Understanding the effects of gravity on time is not only a fascinating component of physics, but it is also important in fields such as satellite technology and space exploration. It reminds us of the complicated interplay of gravity, space, and time, as well as how the fabric of the cosmos impacts our perception of time.

TIME DILATION NEAR MASSIVE OBJECTS

Time dilation near huge objects is an intriguing physics idea derived from Einstein's theory of general relativity. According to this idea, the presence of enormous objects such as planets, stars, or black holes can cause time to flow differently in their vicinity than in areas with lower gravitational effect. In plain terms, time appears to slow down around huge objects in comparison to time in smaller gravitational environments. This phenomenon has been experimentally validated, and it has significant consequences for our understanding of space, time, and gravity.

To understand time dilation near huge objects, we must first understand that gravity is not simply a force that pulls objects towards each other, but rather a curvature of spacetime caused by the existence of mass or energy. This curvature affects the trajectory of things as well as the passage of time. The consequences of time dilation grow more prominent in areas of higher gravity, when spacetime is more severely bent.

The gravitational time dilation experienced by astronauts aboard the International Space Station (ISS) is an example of time dilation near big objects. The International Space Station orbits roughly 408 kilometers above the Earth's surface, where it encounters a somewhat weaker gravitational field than the surface. This difference in gravity causes quantifiable time dilation.

Consider two synchronized clocks, one on Earth's surface and the other aboard the International Space Station. As the seconds pass, it becomes clear that the clock on the ISS is somewhat quicker than the one on the ground. This is because the clock on the surface encounters a larger gravitational force, causing time to move more slowly than the clock on the ISS. This difference in time dilation may appear insignificant, yet it can be measured precisely using precision atomic clocks.

Consider an astronaut spending a year on the International Space Station while his or her twin sibling remains on Earth. Because of the time dilation effect, the astronaut will have aged slightly less than their identical sibling when they return to Earth. This is known as the twin paradox, in which one twin ages less than the other sibling who remains on Earth in a larger gravitational field because one twin experiences weaker gravity during space flight.

This scenario demonstrates the practical ramifications of time dilation in the presence of huge objects. It indicates that time is a pliable entity that may be impacted by gravity rather than an absolute quantity. Experiments employing high-speed particles, satellite-based clocks, and even studies of distant celestial objects have proved the phenomenon of time dilation near huge objects.

In addition to gravitational time dilation, which occurs near big objects, there is another type of time dilation known as velocity time dilation. When an object moves at a substantial fraction of the speed of light, time appears to slow down for it relative to a stationary observer, according to Einstein's theory of special relativity. This effect has been detected in high-speed particle accelerators and validated by several studies.

In summary, time dilation around big objects is a significant result of Einstein's general theory of relativity. It is caused by the bending of spacetime due to the existence of mass or energy. The impact of time dilation is more prominent in areas with higher gravity, when time appears to slow down in comparison to areas with lower gravity. The phenomena has been experimentally validated and has practical ramifications, such as astronauts aboard the International Space Station aging faster than their counterparts on Earth. Time dilation around big objects complicates our understanding of the nature of time, space, and gravity, calling into question our intuitive understanding of time as a constant and universal quantity.

GPS AND THE CORRECTION FOR TIME DILATION

GPS (Global Positioning System) is a satellite-based navigation system that provides users worldwide with precise location and timing information. It uses a network of satellites orbiting the Earth to identify precise positions and enable applications such as navigation, mapping, and clock synchronization. The correction for time dilation, which is required for proper positioning and synchronization, is a critical feature of GPS.

To comprehend the concept of time dilation and its correction in GPS, it is necessary to first grasp the concepts behind GPS operation. The GPS system consists of a constellation of medium-Earth orbiting satellites, each equipped with atomic clocks that give exceedingly precise timing signals. These satellites transmit their signals continuously, which are received by GPS receivers on the ground or in other devices such as cellphones and navigation systems.

Time dilation, a phenomena anticipated by Einstein's theory of relativity, states that gravity and velocity can impact time. In the case of GPS, there are two types of time dilation to consider: gravitational time dilation and relative velocity time dilation.

Gravitational time dilation occurs as a result of gravitational field differences experienced by satellites in orbit and GPS receivers on Earth's surface. The presence of a big object, such as the Earth, creates a gravitational field that can slow down time, according to general relativity. Because satellites are further away from the Earth's gravitational pull than GPS receivers, they have a weaker gravitational field and hence time moves slightly faster for them. This gravitational time dilation effect causes a temporal mismatch between the satellite clocks and the terrestrial clocks.

The relative velocity time dilation is the second type of time dilation in GPS. According to special relativity, when things move at high speeds relative to each other, time appears to flow slower for the moving object when viewed from a stationary frame of reference. In the case of GPS, the satellites move at fast speeds in their orbits relative to the GPS receivers on the ground. As an observer on the ground can see, the relative velocity between the satellites and the receivers causes a time delay.

Both gravitational and relative velocity time dilation effects would produce severe inaccuracies in the GPS system if left untreated. GPS includes a correction system that accounts for time dilation effects to ensure accurate location and timing information.

Adjusting the pace of the satellite's atomic clocks is one method of compensating for gravitational time dilation. To compensate for the slower passage of time in the weaker gravitational field, the clocks on the satellites are designed to tick at a little faster pace than clocks on the ground. This modification keeps the clocks on the satellites and on the ground in sync.

GPS employs the idea of time dilation factor or the relativistic time dilation adjustment to account for relative velocity time dilation. This correction factor is determined by the satellites' velocity in relation to the speed of light and the gravitational potential at their position. By adding this correction factor into GPS calculations, GPS receivers' timing information is corrected to account for the time dilation effect caused by relative velocity.

An example can demonstrate the importance of GPS time dilation correction. Consider a GPS satellite in orbit around 20,000 kilometers above Earth's surface. Because of its altitude, it has a smaller gravitational field than a ground observer. Because of the weaker field, time moves slightly faster for the satellite. Assume the satellite's atomic clock is configured to tick at one second per second as measured on the satellite.

Assume we have a GPS receiver on the ground that is precisely synced with the clock of the satellite at the time of transmission. The signal experiences time dilation as it travels from the satellite to the receiver due to both gravity and relative velocity factors. The receiver's clock would run significantly slower without the correction due to the larger gravitational field and the relative velocity between the satellite and the receiver.

However, the GPS receiver's clock is updated to compensate for these impacts thanks to the time dilation correction process. The correction factor guarantees that the receiver's clock remains in sync with the satellite's clock, accounting for both gravity and relative velocity time dilation. This adjustment enables the receiver to precisely determine the time it took the signal to travel from the satellite, allowing for exact location computations.

To summarize, GPS and time dilation correction are inextricably intertwined. GPS accuracy is dependent on accounting for both gravity and relative velocity time dilation effects. GPS ensures precise location and timing information for users globally by altering the speeds of atomic clocks on satellites and inserting correction factors into calculations. The adjustment for time dilation in GPS shows the practical application of Einstein's theories of relativity and highlights the GPS system's exceptional precision and reliability.

TIME DILATION IN THE UNIVERSE

Time dilation is an intriguing notion in physics that stems from Einstein's theory of relativity. It implies that time is not absolute and can fluctuate depending on observers' relative motion and the strength of gravitational forces. Time dilation demonstrates that the passage of time can be experienced differently for objects under different settings, emphasizing the complex link between time and space.

Time dilation is caused by two basic factors: velocity and gravity. According to special relativity theory, as an object approaches the speed of light, time slows down for that object in comparison to a stationary observer. This effect, known as time dilation owing to velocity, results from the constant speed of light. It indicates that when an object's velocity rises, its internal operations, such as clock ticking, slow down from the perspective of a stationary observer.

Consider the case of a spaceship flying at a considerable fraction of the speed of light to demonstrate the concept of time dilation due to velocity. Assume there are two witnesses, one on Earth and one in the spaceship. The clock on the spaceship appears to be operating slower than the observer's own clock from Earth's perspective. In other words, the traveling clock dilates time relative to the stationary clock on Earth.

However, from the perspective of the observer on the spaceship, their own clock appears to be running regularly, whereas the clock on Earth appears to be ticking slowly. This asymmetry develops as a result of simultaneity relativity, which states that events that are contemporaneous for one observer may not be simultaneous for another moving observer. As a result, both witnesses can properly claim that time is stretched for the other.

As the velocity approaches the speed of light, the time dilation effect gets more dramatic. At such high speeds, time dilation occurs, and an object's internal processes slow drastically in comparison to a stationary observer. Several investigations, including particle accelerators that accelerate subatomic particles to near-light speeds, have proven this occurrence.

Gravity is another component that contributes to time dilation. Gravity warps spacetime, causing clocks to tick slower in areas with larger gravitational fields, according to general relativity theory. This effect, known as gravitational time dilation, suggests that time moves more slowly in the vicinity of enormous objects such as planets, stars, or black holes.

Consider a clock near a big object, such as a black hole, to better comprehend gravitational time dilation. From the perspective of an observer far away from the black hole, the clock near the black hole appears to be running slower than their own clock. This is because the black hole's powerful gravitational pull slows time relative to the observer further away.

The gravitational time dilation effect has also been demonstrated experimentally. Scientists, for example, have employed precision atomic clocks to evaluate time variations between clocks at different elevations. Clocks closer to the Earth's surface, which are subjected to a stronger gravitational field, were discovered to run slightly slower than those at higher elevations.

It is vital to note that the effects of velocity and gravity on time dilation are cumulative. In cases when both velocity and gravity are large, such as near black holes or during high-speed space flight, the combined forces can cause significant time dilation. These effects have practical consequences, particularly for astronauts traveling at relativistic speeds or in close proximity to huge celestial objects.

The famous "twin paradox" is one noteworthy illustration of time dilation's practical ramifications. In this thought experiment, one twin remains on Earth while the other twin travels across space at great speeds, nearing the speed of light. When the traveling twin returns to Earth after a significant amount of time, they will have aged less than their stationary twin. This example demonstrates how velocity-induced time dilation can result in an apparent time difference between observers in different frames of reference.

Finally, time dilation is a fascinating idea derived from Einstein's theory of relativity. It demonstrates that time is not absolute and can fluctuate based on observers' relative motion and the strength of gravitational forces. The passage of time can be experienced differently for objects in different settings due to time dilation caused by velocity and gravity. The phenomena has been empirically confirmed, and it has ramifications for space travel, celestial observations, and our knowledge of the universe. Time dilation calls into question our intuitive understanding of time as a universal constant and pushes us to investigate the complex interplay of time, space, and motion.

PRACTICAL APPLICATIONS OF GRAVITATIONAL TIME DILATION

Gravitational time dilation is an interesting idea in physics that derives from Einstein's general theory of relativity. It implies that gravitational fields can alter time, leading in time dilation or stretching in different regions of space. This phenomena has practical uses ranging from space exploration to commonplace technologies such as GPS devices. In this discussion, we will delve deeper into the practical applications of gravitational time dilation, including a comprehensive example to enhance comprehension.

One practical application of gravitational time dilation is the precise operation of Global Positioning System (GPS) satellites. To identify the position of objects on Earth, GPS uses precise time data. The GPS system is made up of a network of satellites orbiting the Earth, each with atomic clocks that offer incredibly accurate timekeeping. However, due to the effects of gravitational time dilation, clocks on satellites run at a slightly different rate than clocks on Earth's surface.

Because gravitational force is not uniform throughout space, the phenomenon of gravitational time dilation occurs. Greater mass objects, such as the Earth, generate a gravitational field that alters the flow of time. Time travels more slowly in areas with a higher gravitational field, according to general relativity. In the case of GPS satellites in orbit around the Earth, the gravitational field is weaker than that of objects on the planet's surface. As a result, time moves slightly faster for satellites than for terrestrial things.

It is critical to account for the time difference between the satellites and the Earth's surface in order for GPS to function correctly. The GPS calculations would be greatly affected if the effect of gravitational time dilation was not taken into account, resulting in mistakes in estimating the precise position of objects. As a result, the atomic clocks on the satellites are purposefully altered to accommodate for the time dilation effect. To compensate for the differential in gravitational time dilation, clocks on satellites run somewhat quicker than clocks on Earth's surface. This modification enables the GPS system to offer accurate positioning information to users all around the world.

Another practical application of gravitational time dilation can be found in astrophysics, namely the research of black holes. Black holes are extremely dense objects with a powerful gravitational field. Significant time dilation effects are caused by the high gravitational field near a black hole. Time appears to slow down as an object approaches the event horizon of a black hole from the perspective of a distant observer. This is referred to as gravitational time dilation near black holes.

The study of the behavior of light near black holes is a practical application of this principle. Light travelling close to a black hole experiences gravitational redshift, which means its wavelength is stretched due to gravity's effects. As a result, the light appears to have been pushed to the red end of the electromagnetic spectrum.

Scientists can learn a lot about the nature of black holes, their mass, and their gravitational pull on other objects by analyzing this redshift.

Consider a hypothetical situation in which a person travels to a faraway planet with an exceptionally strong gravitational field for a full demonstration of gravitational time dilation. Assume that the gravitational field on this planet is so strong that time dilation effects become apparent. If the individual spends a particular period of time on the planet and then returns to Earth, they will discover that more time has passed on Earth than on the distant world.

For example, suppose the person spends one Earth year on the planet. The effects of gravitational time dilation would have caused the person's clock on the planet to operate slower than clocks on Earth. When they return to Earth, they will discover that several years have passed for humans on Earth, while they have only lived for one year. This highlights the relative nature of time dilation, in which perception of time varies with the strength of the gravitational field.

Finally, gravitational time dilation has practical implications in a variety of domains, including GPS accuracy and the research of black holes. Understanding and accounting for gravitational time dilation effects is critical for accurate computations and observations in these fields. The example of GPS satellites demonstrates the importance of modifying clock rates to account for time dilation. Similarly, research into black holes shows how gravitational time dilation provides vital insights into the behavior of light and the nature of these cosmic objects. Gravitational time dilation illuminates the delicate link between gravity and the flow of time, broadening our grasp of the universe's fundamental workings.

EINSTEIN'S CONSTANT SPEED OF LIGHT

Einstein's constant speed of light, often known as the speed of light in a vacuum, is a basic physics concept that transformed our view of the cosmos. It is represented by the symbol 'c' and has a speed of roughly 299,792,458 meters per second (or approximately 186,282 miles per second). This speed is fundamental in Einstein's theory of relativity, and it has far-reaching implications for our understanding of space, time, and the nature of reality.

One of Einstein's key ideas is that the speed of light is an absolute constant, independent of the motion of the source or the observer. This indicates that no matter how rapidly an observer moves in relation to a light source, the speed of light will always be the same. This contrasts sharply with our ordinary perceptions, in which the speed of objects is determined by their relative motion. If you are in a moving car and throw a ball forward, the ball's speed is the sum of its velocity relative to you and the velocity of the car. However, this is not the case with light.

Consider the following example to demonstrate this notion. Consider yourself to be on a train traveling at a steady pace of 100 kilometers per hour (km/h). A beam of light is thrown forward from within the train. According to classical physics principles, the speed of light emitted by the train should be equal to the sum of the train's velocity (100 km/h) and the speed of light (c). This, however, is not what actually occurs.

In actuality, regardless of the train's velocity, both you and an observer standing outside the train will measure the speed of light to be exactly c. This indicates that regardless of whether the train is stationary or travelling at any constant rate, the speed of light emitted by it will be roughly 299,792,458 meters per second. This is referred to as the constant of the speed of light.

The constant speed of light has far-reaching implications. It indicates that as an object gets closer to the speed of light, its mass appears to grow and time dilation occurs. Einstein's theory of special relativity describes these phenomena. This theory states that as an object accelerates and approaches the speed of light, its mass increases, making it more difficult to accelerate further. To accelerate at the speed of light, an item would need an infinite quantity of energy, which is not achievable in practice.

Time dilation also happens when an object moves at speeds close to the speed of light. Time appears to slow down for a moving object relative to a stationary observer as its velocity approaches c. This means that time moves more slowly for a fast-moving object than for a stationary observer. This effect has been empirically validated and is critical for systems that rely on precise timekeeping, such as GPS.

Light's constant speed has ramifications for our view of space and time. The existence of mass and energy, according to Einstein's general theory of relativity, twists the fabric of spacetime, creating gravitational effects. Gravity may distort the course of light, as illustrated by the gravitational lensing phenomena. One of the primary pieces of evidence supporting general relativity is the bending of light by gravity.

Finally, Einstein's constant speed of light is a fundamental physics idea that claims that the speed of light in a vacuum is always the same, independent of the motion of the source or the observer. This continuous speed has far-reaching ramifications for our concept of space, time, and reality itself. It causes phenomena including time dilation, mass increase at high speeds, and light bending in the presence of gravity. The speed of light's constancy has been scientifically proved, and it serves as the foundation for Einstein's theories of relativity.

TIME DILATION AND THE SPEED OF LIGHT

Time dilation and the speed of light are fascinating concepts derived from Einstein's theory of relativity, which transformed our view of time and its relationship to motion and light. These ideas have far-reaching consequences for our knowledge of the universe, and they have been supported by numerous experiments and observations.

Time dilation is the phenomena in which time appears to pass differently for objects or observers travelling at different speeds or in various gravitational fields. Time dilation happens as a result of the fundamental principle that the speed of light is constant in all inertial frames of reference, regardless of the motion of the source or the observer, according to Einstein's theory of relativity.

Consider the following example to better understand time dilation. Consider two identical twins, Alice and Bob, with Alice remaining on Earth and Bob starting on a high-speed space voyage nearing the speed of light. Before Bob leaves, they synchronize their clocks. As Bob speeds and approaches the speed of light, time appears to pass more slowly for him than it does for Alice on Earth. This means Bob will have aged less than Alice when he returns to Earth after his voyage.

This effect is explained by the fact that the quicker an object moves, the slower time appears to slow for it in comparison to a stationary observer. As the object approaches the speed of light, the effect of time dilation becomes more dramatic. According to relativity theory, when an object's velocity approaches the speed of light, time dilation becomes more intense, eventually resulting in time stopping upon approaching the speed of light.

The speed of light is critical in time dilation. The speed of light in a vacuum, according to Einstein's theory, is constant and is the maximum speed at which information or signals may travel. This unchanging speed of light has far-reaching consequences for our notion of time and motion. The mass of an item grows as it approaches the speed of light, but its length reduces in the direction of motion. These effects, known as length contraction and mass growth, are caused by time dilation.

Because the speed of light is constant, no object with mass can attain or exceed the speed of light. As an object approaches the speed of light, accelerating farther demands an increasing quantity of energy. since of this phenomenon, particles with mass will never reach the speed of light since their energy requirements would become limitless.

Consider the following detailed scenario to demonstrate the notion of time dilation and the speed of light. Consider a spaceship moving near the speed of light in relation to an observer on Earth. A clock on board the spaceship releases a pulse of light every second. If an observer on Earth observes the spaceship travelling at high speed, the light emitted by the clock appears to go a greater distance per second than the stationary clock on Earth.

However, from the perspective of a spacecraft observer, the light emitted by the clock continues to travel at the same speed, c, which is the speed of light in a vacuum. Because the speed of light is constant for all witnesses, the observer on the spacecraft perceives time to move more slowly than the observer on Earth. The spaceship's clock would appear to tick slower than the Earth's clock.

The time dilation effect grows increasingly dramatic as the spaceship approaches the speed of light. Time appears to slow down greatly on the spaceship in comparison to the observer on Earth. If the spaceship travels at a velocity extraordinarily close to the speed of light, the observer on Earth may experience a few years while the occupants of the spaceship experience only a few months. This scenario shows how time dilation becomes significant as an object approaches the speed of light.

Various experiments and observations have confirmed time dilation and the constancy of the speed of light. High-speed particle accelerators, such as the Large Hadron Collider (LHC), rely on time dilation principles to enable accurate measurements and calculations. The GPS (Global Positioning System) also considers the effects of time dilation caused by satellite motion in orbit around the Earth.

Finally, Einstein's theory of relativity relies on time dilation and the constancy of the speed of light. Time dilation is the phenomena in which time appears to pass differently for objects or observers travelling at different speeds or in various gravitational fields. The constant speed of light in a vacuum is important in time dilation because it limits the highest achievable velocity and causes major impacts on time and space when an object approaches the speed of light. These ideas have been experimentally validated, and they have far-reaching consequences for our understanding of time, motion, and the nature of the cosmos.

CONSEQUENCES OF SURPASSING THE SPEED OF LIGHT

Surpassing the speed of light has captivated the imaginations of scientists and science fiction fans alike. According to our current understanding of physics, the speed of light in a vacuum is the universe's ultimate speed limit. However, exceeding this speed would have far-reaching implications for our understanding of fundamental laws and concepts. In this discussion, we will look at the repercussions of exceeding the speed of light and how it affects causality, energy, and our perception of time.

The violation of causality is one of the most serious effects of exceeding the speed of light. Causality refers to the belief that cause and effect relationships occur in a specific order, with the cause coming before the result. This fundamental principle is profoundly embedded in our understanding of the cosmos, allowing us to make predictions and explain natural processes in our surroundings. However, if an object travels faster than light, it may generate situations in which cause and effect get muddled. Effect could come before cause, causing paradoxes and compromising the universe's logical integrity.

Consider the hypothetical case of a superluminal (faster-than-light) spaceship going from Earth to a faraway star system. Assume this spaceship can travel faster than the speed of light and gets to the star system before the light from that system reaches Earth. In this situation, an observer on Earth would see the spaceship arrive before the light signals announcing its departure. This reversal of cause and effect results in the "tachyonic antitelephone," a paradox in which information appears to be conveyed backward in time. Such situations call into doubt the fundamental concepts of causality, as well as the nature of time and the predictability of occurrences.

Another consequence of exceeding the speed of light is that Einstein's theory of relativity, which states that the speed of light is an absolute cosmic speed limit, is violated. This theory states that when an object approaches the speed of light, its mass grows, necessitating an unlimited amount of energy to achieve or exceed that speed. The famous equation E = mc2 captures this concept by demonstrating the equivalence of energy (E) and mass (m) and highlighting the enormous energy required to accelerate an object with mass. Surpassing the speed of light would imply overcoming this limitless energy need, posing a challenge to our knowledge of energy conservation and fundamental physics rules.

Furthermore, the effects of exceeding the speed of light affect our perception of time. Time dilation occurs as an object approaches the speed of light, according to Einstein's theory of relativity. Time dilation is the phenomenon in which time appears to pass more slowly for moving things than for stationary observers. This phenomenon has been empirically validated and is crucial for the operation of GPS satellites, which must account for relativistic time dilation in order to deliver precise positioning data. Surpassing the speed of light would result in even more significant time dilation effects, potentially causing time to appear to flow backward or stop entirely, further challenging our understanding of the nature of time.

Consider the concept of a "tachyon," a hypothetical particle that always travels faster than light, to further demonstrate the repercussions of exceeding the speed of light. Tachyons, if they exist and can be observed, would have unusual qualities. For example, as a tachyon accelerates, its energy reduces rather than increases. This means that tachyons will always travel faster than light, and the more energy they lose, the quicker they will go. Furthermore, a tachyon's movement would be in the opposite direction as its momentum, confounding the accepted link between velocity and momentum. These speculative tachyon properties underscore the peculiar nature of particles traveling faster than light and the significant challenges they pose to our current knowledge of physics.

To summarize, exceeding the speed of light would have far-reaching effects that would call into question our fundamental understanding of the cosmos. It would violate causality, contradict Einstein's theory of relativity, and interfere with our perception of time. While theoretical study of these ramifications broadens our scientific knowledge and piques our interest, it is vital to highlight that there is yet no experimental proof or practical means to achieve superluminal travel. Exploration of these notions helps us to push the frontiers of our understanding, potentially generating new ideas and future discoveries.

INTERSTELLAR TRAVEL AND TIME DILATION

Space, time, and the enormous distances necessary in traveling the cosmos all play a role in the interesting concepts of interstellar travel and time dilation. Further one travels into space, the more noticeable the effects of time dilation become. According to Einstein's theory of relativity, this effect indicates that gravity and velocity can affect the passage of time, leading to discrepancies in how much time has passed in various frames of reference.

Let's look at an example to better grasp how time dilation works in the context of interplanetary travel. Picture a spaceship leaving Earth for a star many, many light-years away. Spacecraft experience temporal dilation as their speeds approach a large percentage of the speed of light. The passage of time within the spaceship will appear to be much slower to an Earth-based observer. This means that while the time visitors spend on the spacecraft may be relatively short, lasting maybe a few years at most, their equivalents on Earth may have aged significantly, possibly by decades or even centuries.

When an object is moving faster than the speed of light or experiencing a greater gravitational field, time passes more slowly for the object than it does for a stationary observer. Time dilation describes this phenomenon. Significant time dilation effects can be brought on by the high velocity obtained by spacecraft during interstellar travel. There will be a large time discrepancy between the space passengers and the stationary spectators as the spacecraft accelerates closer to the speed of light due to time dilation.

Let's think about a made-up example to better grasp this idea. Let's pretend a spaceship with astronauts aboard sets off for a star system 20 light-years away. Their spaceship is optimized for speeds close to the speed of light, allowing them to arrive at their objective in what seems like an acceptable amount of time to them. They set out on their trip, but due to time dilation, they only see a few years pass from their perspective.

On Earth, meantime, decades, if not millennia, had passed. People they knew back then had likely matured or passed away as a result of societal advancements. The astronauts find a world that is very different from the one they left behind when they arrive at their destination. As a result of time dilation, they have effectively journeyed into the future in comparison to Earth.

The extreme consequences of time dilation for interstellar travel are illustrated here. This implies that time could pass very differently for space explorers and Earthlings on long trips to other star systems. Since the travelers' experience of time is different from that of those in other reference frames, the very concept of time becomes subjective and relative.

And as the spaceship approaches the speed of light, the consequences of time dilation will be much more obvious. The time dilation effects become more pronounced the closer the spacecraft's velocity is near the speed of light, leading to increasingly larger discrepancies in the passage of time between the travelers and stationary observers.

It is important to remember that time dilation is not unique to interplanetary travel, but occurs wherever there is a significant change in speed or the presence of a powerful gravitational field. Time dilation, for instance, would be experienced by astronauts circling close to a big object like a black hole because of the strong gravitational attraction.

Time dilation and extrasolar travel are related ideas that stem from Einstein's theory of relativity. As a spaceship approaches the speed of light, time dilation effects become noticeable, creating a time difference between the travelers and stationary observers. The implications of this phenomenon for interstellar travel are intriguing, as explorers to far-flung star systems would only age a few years in their own eyes while decades or centuries pass back on Earth. As time becomes more relative, the concept of time as a linear progression is called into question. The intriguing complication that time dilation brings to future interplanetary travel and the study of the cosmos is something that has always interested me.

THEORETICAL POSSIBILITIES FOR FASTER-THAN-LIGHT TRAVEL

For a long time, the idea of faster-than-light travel has fascinated the human imagination, signifying a science fiction realm that seems beyond the reaches of our knowledge. Although the rules of physics as we know them prevent massive objects from traveling at or faster than the speed of light, many theoretical and speculative frameworks have been proposed to investigate the possibility of overcoming this fundamental limitation. Wormholes, warp engines, and the bending of spacetime are just a few of the out-of-the-ordinary concepts explored by these theories.

The concept of wormholes offers an exciting theoretical option for faster-than-light travel. A wormhole is a hypothetical passage connecting two locations in space and time, allowing travelers to quickly traverse vast distances. Using a wormhole to shorten the time it takes to travel massive intergalactic distances compared to more traditional methods. Wormholes are based on Einstein's general theory of relativity, which postulates that spacetime's curvature can be controlled to generate these tunnels. Wormholes, however, provide formidable difficulties. To prevent the wormhole from collapsing, unusual matter with a negative energy density is needed to stabilize it. It is only conjecture at this point whether or not such exotic substance even exists.

A warp drive is another science fiction-inspired theoretical notion for faster-than-light travel. To achieve faster-than-light travel, a warp drive must alter the fabric of spacetime itself. It is predicated on the idea that a spacecraft can cause a contraction in the area in front of it and an expansion in the area behind it by warping or twisting the fabric of spacetime. A "warp bubble" would form around the spacecraft as a result of this distortion, allowing it to travel faster than the speed of light without violating general relativity. In 1994, Miguel Alcubierre, a theoretical physicist, proposed the first conceptual framework for a warp engine, in which negative energy is used to create a distortion in spacetime. However, an Alcubierre drive would require exotic matter with negative energy densities, like wormholes, which is now beyond anything attainable with existing technology.

Theoretical possibilities for faster-than-light travel should be treated with caution, as they are currently beyond the scope of experimental verification. They ignore quantum mechanical concepts in favor of our current understanding of classical general relativity. Quantum gravity, or some other consistent theory that integrates quantum physics with gravity, may help to clarify the viability of such ideas.

Let's use a made-up scenario to better understand how a warp drive works in theory. Just picture a time when we've made great strides in our understanding of physics and built a real warp drive. In this scenario, a spaceship with a warp drive is sent to investigate uncharted star systems. The warp drive of the ship causes a localized warp in spacetime, enclosing the ship in a sort of warp bubble. The distance the ship needs to go is reduced because the warp bubble compresses space in front of it. Space expands behind the ship, pushing it onward at speeds greater than the speed of light. This allows for quick interstellar travel, as the spacecraft can reach far-off star systems in a matter of weeks or months.

However, the difficulties and restrictions of faster-than-light travel are brought home by the hypothetical example. To create a warp bubble that could travel faster than light would take an enormous amount of energy, which is much beyond the scope of our existing technology and resources. Manipulating spacetime on such a grand scale also has unknown risks and could have far-reaching ramifications for the very fabric of the cosmos.

Conclusion Wormholes and warp drives, two speculative methods for faster-than-light travel, are intriguing ideas that test the limits of our existing physics knowledge. Despite their theoretical foundations in general relativity, they are still very hypothetical and face considerable obstacles, such as the necessity for exotic matter or energy needs that are currently unachievable with present technology. These theoretical concepts may be refined or lead to new breakthroughs as our knowledge of physics grows, potentially ushering in a new era of revolutionary spacetime comprehension and unlocking previously unfathomable possibilities for interstellar travel.

EXPANDING UNIVERSE AND TIME DILATION

Our knowledge of space, time, and the cosmos all contribute to our understanding of the relationship between the expanding universe and time dilation. These ideas are central to contemporary cosmology and have far-reaching consequences for how we conceptualize the expansion of the cosmos and the passage of time.

In astronomy, the term "expanding universe" is used to describe the idea that the distances between galaxies are growing with time. Based on his observations of extremely distant galaxies and their redshift, astronomer Edwin Hubble made this discovery in the early 20th century. Since their light was redshifted, we could tell they were retreating from us at an increasing rate the more away they were. This led to the realization that the universe is dynamic, expanding throughout time.

Time dilation theory is profoundly affected by our expanding cosmos. Einstein's theory of relativity predicts time dilation, which occurs when gravity, velocity, or the relative motion of observers causes time to slow down or speed up. According to the theory of time dilation, the passage of time can be experienced differently depending on the perspective of the observer.

Time slowing occurs as a result of gravity's interaction with the universe's accelerating expansion. General relativity describes gravity as a spacetime curvature resulting from the presence of mass and energy rather than a simple force. Time passes more slowly in locations with a larger gravitational potential, such as those close to huge objects like stars or black holes.

As galaxies recede from one another in an expanding universe, the intergalactic distance increases. This expansion causes a gravitational potential gradient throughout the cosmos, with larger potentials near huge structures like galaxies and galaxy clusters. As a result, these areas experience a slower passage of time compared to those that are not as strongly affected by gravity.

Let's look at an example to better understand this idea. Take Alice and Bob, two witnesses from different parts of the universe. Unlike Bob, who lives in an area with fewer galaxies and less gravitational impact, Alice lives close to a big galaxy. They've synchronized their watches and begun keeping time.

Due to the greater gravitational potential in her neighborhood, time passes more slowly for Alice as she gets closer to the enormous galaxy. Bob, on the other hand, feels time passing more quickly since he is in a location with less gravitational influence. Alice's clock is running slower than Bob's after some time has passed, if they were to compare them.

As the universe continues to expand over billions of years, the time dilation effect caused by the expansion gets increasingly pronounced. Due to differences in gravitational potential brought about by the distribution of matter and the expansion of space, observers in various parts of the universe will experience different rates of time.

The study of cosmology and the development of the universe is affected by the expanding cosmos and time dilation. Evidence for the expansion of the universe can be gleaned from the time dilation seen in distant galaxies, a phenomenon known as cosmic time dilation. It also helps scientists determine the universe's age and see its expansion across time.

As a result of our knowledge of space, time, and gravity, we can conclude that the expansion of the universe and the phenomenon of time dilation are related concepts. Time dilation effects can be seen all around the cosmos because of the differences in gravitational potential caused by the expansion of space. This phenomena, seen in faraway galaxies, is proof that the universe is expanding. To fully grasp the evolution of the cosmos, the passage of time, and the complex interplay between space and time in the cosmos, we must first grasp the concept of time dilation in the context of an expanding universe.

TIME DILATION IN THE EARLY UNIVERSE

As we learn more about the expansion of the universe and the concepts of general relativity, we are presented with the intriguing idea of time dilation in the early cosmos. It's a term used to describe how time seems to move at various speeds throughout the early expanding universe.

About 13.8 billion years ago, the universe was extremely dense and hot, according to the Big Bang theory. Cosmic inflation refers to the time of rapid expansion that occurred as the universe began to expand. Space itself stretched and galaxies moved away from each other as the universe expanded at an exponential rate throughout this inflationary epoch.

Time itself was profoundly affected by the early universe's high levels of energy and density. Curvature of spacetime is predicted by general relativity when mass and energy are present. The consequences of this curvature on time are most pronounced in regions of the cosmos with a higher energy density, such as the early universe.

Time slowing down in the early cosmos can be explained by taking into account the gravitational potential and the impact of cosmic expansion. The gravitational potential fell as the universe expanded fast because the energy density dropped. Time passed more quickly in areas with a lower energy density than in those with a higher density of energy.

Let's use the analogy of two clocks located in separate parts of the early cosmos to clarify this point. One of the clocks is stationed at a place with a high density of matter and energy, where the gravitational potential is greater. As a result of its location in a low-energy-density zone, the gravitational potential of the other clock is reduced.

Time passes more slowly in the high-energy-density zone as the cosmos expands because of the larger gravitational potential, and more quickly in the low-energy-density sector. Gravity and the expansion of spacetime work together to provide this time dilation effect in the early cosmos.

We can learn a lot about the history of the universe and the CMB from accounting for time dilation in the early universe. The cosmic microwave background (CMB) is the Big Bang's lingering radiation, and it sheds light on the early universe. Small temperature changes during the early expansion can be observed by analyzing the CMB. Variations in temperature reflect variations in energy density, and so in the passage of time.

The occurrence of time dilation in the early cosmos has been established by CMB studies. The variations in energy density and gravitational potential are reflected in the temperature differences seen across the CMB map. Scientists can learn more about the mechanics of the early cosmos, its expansion, and the distribution of matter and energy by investigating these temperature oscillations.

As a result of the early universe's rapid expansion and extremely high energy density, time slowed down. This causes time to move at different speeds in various parts of the universe, with higher energy density regions experiencing a slower flow of time. To fully grasp the dynamics of cosmic expansion, the emergence of structures, and the development of our universe from its inception to its present condition, an appreciation of time dilation in the early cosmos is essential.

THE CONCEPT OF COSMIC TIME

Understanding cosmic time is crucial to comprehending the development of the cosmos. It's a way of thinking about time that spans cosmic dimensions, from the Big Bang to the current day. Here, we'll investigate the meaning, measurement, and connection of cosmic time to the expansion of the universe. The observation of distant galaxies and the idea of lookback time provides an intriguing case study that highlights the tremendous ramifications of cosmic time.

The scientific study of the cosmos relies heavily on the concept of cosmic time. It enables us to observe the development of phenomena and events over extremely long periods of time. Cosmic time spans billions of years, from the birth of stars and the assembly of galaxies to the expansion of the universe itself, which is a far cry from the human perspective of time.

In terms of cosmic history, the expansion of the universe is a central idea. The Big Bang theory postulates that roughly 13.8 billion years ago, the cosmos began in a very hot and dense condition. It has been growing ever since, with galaxies receding from one another. The ramifications of this growth for our ability to measure and comprehend the passage of cosmic time are substantial.

Scientists use a wide range of methods and data to determine cosmic epochs. The redshift of light from extremely distant galaxies is one of the primary instruments used. The light emitted by galaxies has a longer wavelength, or "redder" light, because of the expansion of the universe. Scientists can roughly calculate the amount of cosmic time that has passed since the light was emitted by measuring the magnitude of this redshift. This gives us a powerful tool for dating distant astronomical phenomena and objects.

Let's take a closer look at an intriguing example that illustrates cosmic time: the observation of distant galaxies and the idea of lookback time. Galaxies at tremendous distances from us are like staring into the past when we study them. The light we receive from distant galaxies has taken millions or possibly billions of years to traverse the vastness of space to reach us.

For example, if we look at a galaxy that is a billion light-years away, we are actually viewing it as it looked a billion years ago. That's what we call "lookback time" in the biz. When we expand our view of the universe, we can also expand our view of the past. Astronomers can learn about the universe's infancy, origin, and development through the study of these faraway galaxies.

Amazing results have been found in case studies of faraway galaxies. The expansion of the universe and the presence of dark energy, for instance, have been corroborated by studies of distant galaxies separated by billions of light-years. The analysis of this galaxy's light has led scientists to conclude that the expansion of the universe is speeding, pointing to the presence of a mysterious component called dark energy. This finding has significant ramifications for how we see the progression of the cosmos and the passage of cosmic time.

In addition, we can learn more about the cosmic microwave

background (CMB) radiation by observing distant galaxies. The cosmic microwave background (CMB) is the diffuse afterglow of the Big Bang. The cosmic microwave background (CMB) allows scientists to peer back in time to the universe's infancy, when it was just a few hundred thousand years old, by studying its characteristics. These studies help us better understand the density and temperature of the universe at various epochs in its history.

To sum up, the idea of cosmic time is a cornerstone of cosmology because it allows us to examine the development and history of the universe throughout enormous intervals of time. It includes everything from the birth of galaxies to the death of stars to the expansion of the cosmos. Scientists can determine the ages of cosmic objects and events by using methods like detecting redshift and observing faraway galaxies. The observation of distant galaxies and the concept of lookback time serve as an excellent example of the far-reaching consequences of cosmic time, providing new understanding of the early universe, the acceleration of the expansion of the cosmos, and the existence of dark energy. Understanding the vastness and complexity of the ever-changing cosmos is greatly enhanced by looking into its past.

THE ROLE OF TIME DILATION IN THE FATE OF THE UNIVERSE

Time dilation is a key idea for making sense of the future of the universe. Time dilation occurs when gravity or high speeds cause the passage of time to be slowed down or lengthened. It has far-reaching consequences for our understanding of cosmological time, the expansion pace of the cosmos, and the ultimate fate of celestial objects. In this investigation, we will delve into the significance of time dilation in the ultimate fate of the universe by looking at how it influences different cosmic events and what it means for our understanding of the cosmos' ultimate fate.

Gravity is a major contributor to time slowing down. Einstein's general theory of relativity predicts that large masses, like stars and black holes, can distort spacetime and slow the passage of time in their immediate vicinity. Numerous tests and observations have proven this phenomena, known as gravitational time dilation. For instance, because to changes in gravitational forces, atomic clocks located at various heights or in close proximity to enormous objects tick at very different rates.

When pondering the fate of the cosmos, gravity and time dilation take on added importance. It is generally accepted that the universe is, and will continue to be, expanding. The outcome of this expansion, however, will be determined by the total mass and energy of the cosmos, as well as the critical density necessary to halt the expansion. What happens to celestial objects and the universe as a whole is ultimately determined by the interplay between gravity, time dilation, and the expansion rate.

The universe will expand forever if the total mass and energy it contains are below the critical density. Time dilation plays a part in determining the fate of celestial objects in this scenario. As the universe expands, the distances between galaxies and galaxy clusters increase. It slows down processes like star formation and stellar evolution because less gravitational interactions are being generated between the objects involved. Time dilation would alter the general rate of cosmic events, such as the birth and death of stars, slowing them down.

However, if the density of the universe's matter and energy is too high, gravity will win out and the expansion will reverse, resulting in the Big Crunch. In this case, the timescale of the collapse is affected by time dilation. The consequences of time dilation intensify when the universe contracts under the pull of gravity. From the outside looking in, events within collapsing structures would appear to happen faster, such as the death of stars and the merger of galaxies.

It's worth noting that scientists are continuously investigating and debating the fate of the cosmos. The amount of matter and energy in the universe can be inferred from observations of the cosmic microwave background radiation, the distribution of matter in the cosmos, and the expansion rate. However, the fate of the universe and the function of time dilation may also be affected by other, as-yet-ununderstood forces, such as dark matter and dark energy.

Let's think about a made-up scenario to demonstrate the role time dilation plays in the fate of the cosmos. Consider an extremely faraway galaxy that is rapidly receding from our own. Time for things within the galaxy would slow down with respect to a stationary observer, as predicted by the theory of special relativity. So, from our vantage point, internal activities within that galaxy, like star formation and the evolution of stellar populations, would appear to take place at a slower rate.

Now, if we skip ahead a few billion years, the universe is still expanding, and the distance between our observer and the faraway galaxy keeps growing. Time dilation is amplified by the cosmos' expansion and the speed at which the galaxy is moving. Due to the accelerating expansion of the universe, the time experienced by things within the galaxy appears to move even more slowly from our vantage point.

The fate of celestial objects and the development of the universe as a whole can be profoundly affected by time dilation, which is influenced by both velocity and cosmic expansion. This shows how gravity, expansion, and time dilation all interact to determine the fate of the universe.

In conclusion, time dilation is an important factor in figuring out how the cosmos will end up. From star formation and evolution to the expansion of the cosmos, its effects can be felt everywhere. Whether the universe expands forever or has a reversal known as the Big Crunch is determined by the interplay between gravity, the critical density, and the amount of matter and energy in the cosmos. The idea of time dilation helps us better comprehend the ultimate fate of the universe by shedding light on the timeframes and dynamics of major cosmic processes.

TIME DILATION AND THE COSMIC MICROWAVE BACKGROUND

Two fascinating cosmological notions that shed light on the cosmos and its early stages are time dilation and the cosmic microwave background (CMB). Time dilation, predicted by Einstein's theory of relativity, occurs when the passage of time is slowed or sped up as a result of external influences like gravity or high velocity. In contrast, the ubiquitous radiation known as the cosmic microwave background is a relic from the early cosmos that allows us to peer back in time and see how things were back then. In this investigation, we will look into the connection between time dilation and the cosmic microwave background, analyzing their interplay and the insights it might provide about the cosmos.

Let's look into the history of the cosmic microwave background (CMB) to see how time dilation fits into it. After the universe had been for some 380,000 years following the Big Bang, it had cooled down to the point where atoms could form. Before this time, space was filled with a hot, dense plasma of particles and radiation. Photons, the particles that make up light, became unbound from matter as the cosmos grew and cooled. Over billions of years, the primordial photons, which were originally high-energy radiation, were stretched and cooled, transforming them into microwave radiation. Today, we call this pervasive but chilled radiation the cosmic microwave background.

The role that time dilation plays in the context of the cosmic microwave background will now be investigated. When spacetime is stretched or warped due to factors like gravity or very high velocities, time slows down as a result. The universe was significantly denser and hotter when the cosmic microwave background was emitted than it is now. Gravitational forces were noticeable, and time dilation was noticeable, at such high densities and temperatures.

Cosmic inflation, the fast expansion of the cosmos, occurred during the time when the cosmic microwave background was emitted. During this inflationary period, spacetime stretched and expanded at an unprecedented rate. This resulted in time dilation for the photons that make up the cosmic microwave background, so that their ticking speed appeared to slow down in comparison to that of a stationary observer. This means that the photons' passage from the place of emission to the present has been lengthened, according to the experience of time by a stationary observer.

Our ability to observe and detect the CMB is profoundly affected by the time dilation phenomenon that occurs during CMB emission. CMB photons have been subject to diverse gravitational fields and expansion rates as they have traversed the expanding cosmos and passed through different areas. Consequently, the apparent frequencies and wavelengths are modified by the varying degrees of time dilation experienced by the photons along their path.

Variations in the apparent temperature of the cosmic microwave background can be traced back to these localized time-dilation effects. The photons' extended journey through a variety of gravitational fields and expansion rates causes them to exhibit small temperature fluctuations, which are mirrored in the CMB. Accurate temperature data are studied by scientists to learn more about the early universe's structure and composition, as well as the physical processes that drove its evolution.

The cosmic microwave background radiation map is one of the most useful tools for investigating the cosmic microwave background. This map shows the distribution of CMB temperature variations over the sky in great detail. Scientists can learn a lot about the universe's make-up, the existence of dark matter and dark energy, and the conditions that led to the development of galaxies and other large-scale structures by investigating the patterns and distribution of these fluctuations.

In conclusion, there is a close relationship between time dilation and the cosmic microwave background. Time dilation effects on the photons that make up the CMB are significant since the cosmos was expanding rapidly when the CMB was being emitted. These temperature oscillations in the CMB can be attributed, in part, to time dilation effects brought on by the stretching and warping of spacetime. Scientists can learn a lot about the early cosmos, its makeup, and the physical processes that influenced its formation by analyzing these variations. Understanding the origins of the universe and the significance of the information provided by the cosmic microwave background is made possible by this extraordinary window into the early phases of the universe.

TIME DILATION IN PARTICLE ACCELERATORS

Einstein's theory of relativity gives rise to a fascinating phenomena called time dilation, which has far-reaching ramifications in many branches of science. Particle accelerators, where scientists investigate elementary particles and the fundamental forces that hold the cosmos together, are one area where temporal dilation plays a key role. To better comprehend the fundamental building blocks of the universe, we will delve into the notion of time dilation in particle accelerators, how it influences the behavior of particles, and its practical ramifications.

First, some groundwork is required to grasp the concept of time dilation in particle accelerators. Time, in Einstein's theory of relativity, is not an absolute but rather depends on the perspective of the observer. Time slows down for an object traveling close to the speed of light, according to the idea. Time dilation is the term for this occurrence. When particles are accelerated to near the speed of light in particle accelerators, they encounter considerable time dilation effects.

Large equipment called particle accelerators are used to boost the kinetic energy of charged particles like protons and electrons to unprecedented levels. Scientists can then investigate the interactions and particles produced by these colliding or guided beams of accelerating particles. The underlying nature of matter, the properties of particles, and the fundamental forces that govern them can be explored by recreating high-energy particle collisions in controlled laboratory environments.

Time dilation plays a significant role in particle accelerators because the accelerated particles reach speeds near to the speed of light. Compared to a stationary observer, the particles' clocks or internal operations will slow down as they approach the speed of light. Therefore, particles see time differently from a stationary observer.

Particle behavior and scientific observations are both affected by time dilation's presence in particle accelerators. From the vantage point of a stationary observer, time dilation has the effect of lengthening the "lifetime" or "decay time" of the particles. This is because their fast speeds cause their clocks to tick more slowly.

Take the muon, for instance, which has a well-established average lifetime. After a certain average lifespan, muons decay into other particles. A particle's "apparent" lifetime increases due to time dilation effects as it is accelerated to high velocities in a particle accelerator. Scientists are able to investigate muon behavior and conduct experiments that would be impossible without the time dilation effect because of this increased lifetime.

The measurements scientists take in particle accelerators are also affected by time dilation. For a deeper understanding of matter and the processes that shape it, scientists rely on accurate measurements of particle attributes like mass, energy, and decay rates. However, time dilation effects require a correction to these observations to account for the different rates at which the accelerated particles and the stationary observer's clocks are ticking.

For precise measurements and reliable interpretations of experimental results, time-dilation corrections are necessary. Scientists use high-tech methods and mathematical calculations based on relativity theory to make these adjustments. The effects of time dilation can be taken into account in particle characteristics determination, theoretical model verification, and cosmological frontier research.

Particle accelerators have made tremendous discoveries and advancements in particle physics thanks to the study of time dilation. For instance, the Higgs boson, a particle essential to the mechanism of mass creation, has been experimentally verified to exist by researchers using the Large Hadron Collider (LHC) at CERN, the European Organization for Nuclear Research. These ground-breaking findings owe a debt of gratitude to the exact measurements of particle characteristics, which included the effects of temporal dilation.

Time dilation, a key notion in Einstein's theory of relativity, is thus crucial to the study of particle accelerators. It's a crucial piece of the puzzle when trying to figure out what happens to particles when they're propelled to nearly the speed of light. Particles' apparent lifetimes are altered by time dilation, allowing researchers to probe their properties and conduct studies that would be impossible without it. As a result, adjustments must be made to data to account for the time discrepancy between the accelerated particles and the stationary observers. Particle accelerator research on time dilation has yielded important insights into the nature of the universe's basic blocks.

TIME DILATION IN ATOMIC CLOCKS

The Einsteinian phenomena of time dilation is fascinating and has important implications in many branches of science and industry. Atomic clocks, which measure time with extreme precision thanks to the way in which individual atoms behave, are one area where time dilation plays an important role. We'll look at what time dilation is, how it impacts atomic clocks' ability to keep accurate time, and what it means for modern life.

To provide extremely accurate timekeeping, atomic clocks rely on the natural frequency of atomic vibrations or oscillations. Atomic features, such as resonant frequencies, are used to provide a reliable and accurate time standard. The cesium atomic clock is the most popular form of atomic clock because it uses vibrations of cesium atoms to define the second.

We must go into the concepts of relativity to comprehend time dilation in atomic clocks. Time, in Einstein's theory of relativity, is not an absolute but rather depends on the perspective of the observer. According to the hypothesis, as an object approaches speeds close to the speed of light, time slows down for the object in comparison to a stationary observer. Time dilation is the term for this occurrence.

Time dilation becomes significant for atomic clocks when they are in motion or when they are subjected to strong gravitational fields. Both velocity time dilation and gravity time dilation play a significant role in the time distortions measured by atomic clocks.

When an atomic clock is traveling at great speeds in comparison to a stationary observer, a phenomenon known as velocity time dilation takes place. When a clock moves closer and closer to the speed of light, the atomic vibrations that keep it running slow down, as predicted by relativity. From a stationary vantage point, the passage of time thus appears to slow down.

However, when gravity affects the passage of time, a phenomenon known as gravitational time dilation takes place. A stronger gravitational field, according to relativity, causes clocks to tick more slowly than those in a weaker field. This means that an atomic clock located on the surface of the Earth will experience a slower rate of time than one located at a higher altitude because of the gravitational pull of the planet.

Atomic clocks are affected by both the effects of speed and gravity on the passage of time. For example, GPS systems rely on accurate timekeeping to pinpoint their locations. Because of their orbital velocities and their altitudes in Earth's gravitational field, the atomic clocks onboard GPS satellites experience both velocity time dilation and gravitational time dilation. The precision of GPS locating would suffer greatly if the effects of time dilation weren't taken into account.

Scientists and engineers make adjustments to the readings of atomic clocks to account for time dilation effects. By making these adjustments, atomic clocks may keep their time readings in sync with the reference time kept by bodies like the International Bureau of Weights and Measures (BIPM).

Time dilation corrections are based on relativity theory and necessitate accurate measurements of the clock's velocity and the local gravitational field strength. By making these adjustments, scientists can keep the global network of atomic clocks in perfect sync with one another.

Time dilation research on atomic clocks has led to significant improvements in both timekeeping and science. Accurate measurements in industries like navigation, telecommunications, and basic science have benefited from the increased precision and stability of atomic clocks.

In the realm of fundamental physics, atomic clocks are put to use in the testing of relativity principles and the investigation of possible violations of these theories. Atomic clocks allow scientists to test the bounds of our knowledge of the universe and look for phenomena beyond the Standard Model by comparing their performance in different gravitational fields and under varied velocities.

In conclusion, atomic clocks, which are extremely precise timekeeping devices based on the behavior of atoms, rely heavily on the phenomenon of time dilation. Time kept by atomic clocks must be adjusted to account for the effects of both velocity and gravitational time dilation. Time dilation research has helped develop fundamental physics and has real-world implications in areas like GPS navigation and communications. The precision of atomic clocks and our knowledge of time itself are both expanding as a result of continual scientific and technical developments.

ADVANCEMENTS IN TIME MEASUREMENT

Improvements in the ability to measure time have played a pivotal role throughout human history, allowing us to better comprehend and manage the environment. Timekeeping has come a long way from crude sundials to modern, accurate atomic clocks thanks to several scientific discoveries and technical developments. In this investigation, we'll look into the evolution of timekeeping technology, highlighting major developments and the ways in which they influenced other disciplines like astronomy and GPS.

Natural events and astronomical observations were the basis of early clocks. One of the earliest methods of telling time was with a sundial, which measured the angle of the shadow cast by the sun to determine the time of day. Taking into account seasonal changes and geographical settings, ancient cultures like the Egyptians and the Greeks created sophisticated sundials. The regulated flow of water was used to mark the passage of time in the water clock, also known as a clepsydra, which was an improvement over previous methods of measuring time.

Measuring time was much improved with the introduction of mechanical clocks in medieval Europe. Weighted or spring-driven mechanical clocks were a more reliable and transportable way to keep track of time. These early clocks weren't very accurate until escapement mechanisms like the verge escapement and the pendulum were invented. As cultures came to rely on accurate timekeeping for making forecasts, planning nautical expeditions, and organizing economic operations, the invention of mechanical clocks had far-reaching effects.

Timekeeping improved even more during the 18th and 19th centuries as a result of the Industrial Revolution. John Harrison's creation of the chronometer revolutionized maritime navigation by making precise timekeeping on extended trips possible. With its balancing wheel and spring system, Harrison's chronometer was a compact and accurate way to keep track of time on the high seas. This new development helped significantly with oceanographic surveys.

In the 20th century, thanks to scientific and technological progress, clock accuracy reached previously unimaginable heights. The advent of atomic clocks made possible by the study of subatomic particles was a game-changer in the history of horology. Atomic clocks calculate time with extreme precision by tracking the vibrations of atoms, most commonly cesium but also rubidium and ytterbium. According to the International System of Units (SI), one second is equal to the time it takes for the cesium atom to go from one energy level to the next. This is expressed as 9,192,631,770 radiation cycles. Since then, atomic clocks have replaced all other types of timepieces, improving precision in the fields of science, communications, and navigation around the world.

Another important step forward in timekeeping was the advent of quartz crystal clocks in the middle of the twentieth century. The piezoelectric properties of quartz are harnessed in quartz clocks to produce extremely stable electrical oscillations. These clocks, which may be found in things like wristwatches and electronic devices, are more widely available to the public because of their lower cost and portability in comparison to atomic clocks.

Recent developments in optical and atomic physics have enabled greater precision in measuring time. With their ability to produce femtosecond (one quadrillionth of a second) ultra-short pulses, femtosecond lasers have enabled novel avenues of investigation and technological advancement. Using these lasers in tandem with cutting-edge measurement techniques, researchers have made significant strides in their knowledge of fundamental physical phenomena by studying ultrafast processes at the atomic and molecular levels.

Another noteworthy improvement in chronometry is the creation of satellite-based navigation systems like GPS. Accurate positioning and navigation anywhere in the world is made possible thanks to GPS, which uses atomic clocks on board satellites to send out extremely accurate time signals. Accurate geolocation requires the synchronization of these atomic clocks and the accurate assessment of signal transit durations.

Improvements in the ability to measure time have had far-reaching consequences across a range of scientific fields and practical contexts. Accurate timekeeping is essential for the observation of astronomical events, the study of planetary motions, and the detection of gravitational waves. Accurate time synchronization is essential for the smooth operation of communications networks and the transfer of data. Countless other disciplines rely heavily on accurate timekeeping, including finance, meteorology, geology, and many more.

Sundials, water clocks, and atomic clocks have all given way to femtosecond lasers and other cutting-edge technologies in the quest to accurately measure time. Each achievement has pushed the limits of precision and accuracy, allowing us to better understand the world and make strides in our daily lives. Scientific research, technological progress, and the search for more precise and dependable timekeeping systems are all driven, in part, by the evolution of new methods for measuring time. Our navigational, linguistic, and geographical abilities improve together with our knowledge of time.

TIME DILATION IN QUANTUM MECHANICS

Einstein's theory of relativity, which transformed our understanding of space, time, and gravity, gives rise to the interesting concept of time dilation. The concept of time dilation is most often linked with the macroscopic world of general relativity, but it also plays a key role in quantum mechanics, the area of physics that studies the behavior of particles on the tiniest sizes. The importance of time dilation to our knowledge of quantum mechanics and the quantum world will be discussed, along with its consequences for particle behavior and some concrete instances.

Particles are characterized by wavefunctions in quantum mechanics, and these wavefunctions change over time according to a mathematical equation called the Schrödinger equation. The probabilistic behavior of particles is governed by this equation, which allows us to determine the likelihood of observing a particle in a given state at a given instant. However, the effects of relativistic motion and gravity can dramatically alter the behavior of particles, and this must be taken into account when discussing time dilation in the quantum realm.

High-energy particle collisions are one scenario where time dilation plays a crucial role in quantum mechanics. Colliding particles at particle accelerators like CERN's Large Hadron Collider (LHC) are propelled to nearly the speed of light. These high-energy events, produced by the collisions, are what scientists use to learn about the building blocks of matter and the forces that govern it. However, time dilation influences the time evolution of particles' wavefunctions as they approach the speed of light.

As a result of the time dilation effect, particles involved in these high-energy collisions see time differently than a stationary observer. This means that the wavefunctions, which describe the particles' internal operations, change more slowly as seen by a stationary observer. Particles' dynamics, decay rates, and the probabilities attached to various events are all consequently modified by time dilation.

As an illustration, think about the process by which a muon, a highly energetic unstable particle, created in a particle collision, decays. The typical amount of time it takes for a muon to decay is well-established. When muons are driven to very high speeds, however, their apparent lifetime increases due to time dilation. This means that the muons appear to have a longer half-life from the vantage point of a stationary observer. This effect on time enables for more thorough investigation of muon properties and behavior than would be feasible without time dilation.

Particle lifetimes are measured by particle detectors, and time dilation plays a critical role in this process. High-energy collisions generate particles, which can be detected and studied with particle detectors. Particle lifetimes are determined by studying the times at which particles hit the detector. On the other hand, due to time dilation, particles traveling at fast speeds have a later arrival time relative to the detector's clock.

Scientists need to make adjustments when interpreting

experimental data to account for the effects of time dilation. The measured lifetimes are then corrected to account for the particles' rest frame features and decay rates. Scientists can verify the predictions of quantum mechanics by obtaining exact measurements after accounting for temporal dilation.

Extreme gravitational conditions, such as those found close to black holes, further magnify the significance of time dilation in quantum physics. According to general relativity, a huge object, such as a black hole, will warp spacetime due to the presence of its gravitational field. As a result of this distortion, time slows down significantly for an observer in the area of the black hole compared to an observer further away.

Time dilation close to a black hole can have significant ramifications in quantum mechanics. The effects of time dilation around the event horizon of a black hole can be seen, for instance, in the phenomena of Hawking radiation, which predicts that black holes leak radiation over time. The quantum fluctuations of particle-antiparticle pairs near the black hole are affected by the apparently slow passage of time at the event horizon, resulting in the production of Hawking radiation.

To sum up, time dilation is extremely important in quantum mechanics. Particle lifetime measurements and the evolution of wavefunctions in high-energy collisions are just two examples of how time dilation might alter the odds of an encounter. The importance of time dilation in comprehending the quantum universe is demonstrated by examples such as the increased lifetime of speeding particles and the corrections employed in experimental measurements. Furthermore, the strong impact of time dilation on phenomena like Hawking radiation is shown at the extreme gravitational circumstances surrounding black holes. Scientists can improve their grasp of quantum physics and delve further into the composition and interactions of subatomic particles by factoring in the effects of time dilation.

TIME TRAVEL AND TIME DILATION

The ideas of time travel and time dilation are fascinating and have sparked the interest of many. Although time travel is still science fiction, Albert Einstein's theory of relativity sheds light on the connections between space, time, and motion. According to this theory, external forces like velocity and gravity can cause time to slow down, a phenomenon known as time dilation. In this investigation, we will examine the fascinating implications of time travel and time dilation, as well as their relationship to one another and to a case study.

The theory of relativity predicts that the passage of time will be different for two observers who are traveling relative to one another or who are in distinct gravitational fields. This phenomenon is known as time dilation. Time is said to pass more slowly for moving objects and in greater gravitational fields than it does for stationary ones.

The "twin paradox" is a well-known thought experiment that can help us make sense of time dilation. Picture Alice and Bob, two conjoined twins. While Alice stays on Earth, Bob travels into space at nearly the speed of light. After some time away from Earth, Bob will have aged more slowly than Alice when he eventually returns. Bob's rapid velocity in comparison to Alice's caused a dilation of time.

Time seemed to be passing normally inside the spaceship from Bob's vantage point, but Alice's time on Earth seemed to be passing more slowly than usual due to Bob's high velocity. The theory of relativity foretold this time dilation effect. It is important to note that both Alice and Bob consider their own experience of time to be normal, but that when contrasted to the other's, the differences become more apparent.

The interplay between gravity and time dilation is also crucial. Whether it's a planet or a black hole, the presence of a huge object warps spacetime in accordance with general relativity. The gravitational field produced by this curvature modifies the flow of time. Time passes more slowly the closer you are to the gravitational center compared to further away. The experimental confirmation of this gravitational time dilation is essential for the precision of GPS systems.

Time dilation is a phenomena with far-reaching ramifications for the idea of time travel. Time travel into the future is a scientific possibility, but back into the past is still up for debate and is currently only a theory. Fiction and theoretical physics have both investigated time dilation and wormholes, which are theoretical shortcuts in spacetime.

Time dilation provides the potential for perceiving time at a slower rate compared to the external world in the event of future time travel. This is possible by entering places with strong gravitational fields or by traveling at speeds close to the speed of light. Therefore, a person taking such a trip might find that considerable time has passed in the external world while they themselves have aged very little.

The concept of retrograde time travel, or time travel in the past, offers considerable difficulties and contradictions. Fundamental problems about the viability of time travel arise from the concept of causality, which holds that an event cannot precede its cause. Further complicating the idea of retrograde time travel is the prevalence of paradoxes like the grandfather paradox, in which one might theoretically prevent their own existence.

It is important to separate the scientific hypotheses about time travel from the imaginative depictions of time travel in popular culture, even though the concept of time travel remains a matter of scientific and philosophical inquiry. Time, space, and the feasibility of time travel are all topics that scientists are still investigating and theorizing about.

Time dilation, a result of relativity theory, exemplifies the interesting interplay amongst time, motion, and gravity. Time dilation and the possibility of time travel have piqued curiosity among humans for generations. While time dilation has been empirically proved and is important to our understanding of the cosmos, backwards travel in time is still a matter of conjecture at best. There are fascinating new directions that research, philosophy, and the arts can take thanks to the interplay between time dilation and time travel.

IMPLICATIONS OF TIME DILATION ON HUMAN PERCEPTION

Human perception is profoundly affected by time dilation, a central concept in relativity theory. Time dilation results from the interplay between the passage of time and other factors, such as velocity and gravity. Time dilation is a verified scientific fact, but the way it affects people's perceptions of time is fascinating. Time dilation has profound ramifications for how we age, think, and experience the here and now, all of which we'll examine in this investigation.

The effects of temporal dilation on the aging process are particularly noteworthy. Time slows down for a moving object as it approaches the speed of light or travels through strong gravitational fields, as predicted by the theory of relativity. This indicates that people in strong gravitational fields or while in motion age more slowly than people in a different reference frame. The effects of time dilation on human aging are small under normal conditions, but they become considerable when traveling at very high speeds or when near extremely huge objects like black holes.

The time dilation effect on human aging is best demonstrated by the "twin paradox." In this hypothetical situation, one twin travels at light speed through space while their identical sibling stays on

Earth. Time dilation means that the twin who has been traveling will have aged less than his or her counterpart who has remained in one place. This phenomena demonstrates the relative nature of aging and complicates our intuitive notion of time. It prompts contemplation on the relative nature of our experiences and the ways in which observers in different reference frames perceive time differently.

The cognitive and perceptual processes are also affected by time dilation. The human brain uses temporal processing to make sense of events and construct meaningful stories from them. But in situations that cause time dilation, such moving at high speeds or being in strong gravitational fields, our internal clocks may not sync up with the clocks around us.

A person's internal sense of time is unaffected by time dilation. From the outside, though, it may seem as though time is moving more slowly for them. This disconnect between our internal and exterior experiences of time raises questions about the objectivity of time and might cause cognitive discomfort. It prompts thought-provoking queries on the place of individual perception in the formation of objective reality.

In addition, the effects of temporal dilation complicate our standard conception of the here and now. In common experience, the present is experienced as transitory and instantaneous. Time dilation, however, adds a new dimension of relativity to our understanding of the present. When people see time differently, their perceptions of the present can change.

One notion that exemplifies this idea is "relativity of simultaneity." This theory states that the simultaneity of events might change as a result of the motion of the observers. A second observer in a separate reference frame could see two occurrences that the first sees as happening at the same time as happening at different times. This perspective on simultaneity, which is grounded on relativity theory, forces us to reevaluate our assumptions about the absolute nature of time.

The consequences of time dilation on human perception are a reminder of the fluidity of time and the complexity of the interplay between these three dimensions of spacetime and the forces of gravity and motion. They force us to reevaluate our preconceived notions of reality and force us to reconsider our intuitive knowledge of time as an absolute and universal entity.

In sum, the effects of time dilation on our perception are far-reaching. Its implications for how we age, how we think, and how we make sense of the here and now make us ponder the subjectivity of time and the nature of our own experiences. The idea of time dilation calls into question the static and universal character of our conception of time and instead highlights its inherent fluidity and relativity. The effects of time dilation on human perception have deep ramifications that challenge us to ponder the secrets of time and to broaden our perspective on the cosmos.

THE PHILOSOPHICAL IMPLICATIONS OF TIME DILATION

The philosophical ramifications of time dilation, a concept derived from relativity theory, are significant enough to call into question our fundamental assumptions about the nature of time, reality, and even our very existence. Time dilation prompts philosophical investigation into the nature of subjective experience, the connection between time and consciousness, and the essence of reality by casting doubt on the absoluteness and linearity of time.

Time's relativity is one of the most important philosophical consequences of time dilation. According to relativity theory, the rate at which time passes changes based on the locations and speeds of observers, as well as the strength of their gravitational fields. The idea that time is a fixed, unchanging entity is thus called into question. Instead, time becomes a personal sensation that depends heavily on the perspective of the observer.

Time relativity poses serious challenges to our understanding of the world and our place in it. The idea that time is intrinsically linked to our subjective experience rather than existing apart from it is put forth. This runs counter to the widely held belief that there exists a timeless, objective reality. Questions over whether or not time is a basic part of the fabric of the cosmos or merely a creation of human experience arise as a result.

The experience of time slowing down causes one to ponder profound questions about life and the self. Because of the close relationship between time and our awareness, the fact that different people perceive time at various speeds raises questions about the stability and indivisibility of the individual. Time relativity casts doubt on the idea of a fixed, universal self moving through time in a straight line. Instead, it hints at a splintering of subjective realities due to the possibility that various observers experience time differently.

Intriguing metaphysical concerns regarding the nature of reality outside of our subjective experience are also raised by time dilation. We need to reevaluate the ontological position of time if it is flexible and affected by external forces like motion and gravity. Is time something that exists apart from our awareness of it, or do we create it through our interaction with the world?

Time dilation has far-reaching philosophical consequences, including discussions of the arrow of time, causation, and the nature of change. Since events may appear to have occurred in a different temporal order to various observers, the subjective nature of time complicates our ability to grasp the connections between causes and effects. As the relativity of simultaneity doubts the existence of a single "now" shared by all observers, it also begs questions about the directionality of time.

We are led to ponder the fundamental nature of time, the role of consciousness in moulding our temporal experience, and the nature of reality itself as we consider the philosophical implications of time dilation. Time dilation forces us to rethink our assumptions, opening us up to the world's intrinsic intricacies and mysteries.

In conclusion, the philosophical ramifications of time dilation pose serious problems for our common sense interpretation of how the world works. The nature of subjective experience, the role of time in consciousness, and the relativity of time are all called into doubt. Time's relativity provokes metaphysical inquiries into its ontological character, as well as contemplation of the essence of existence and the self. Questions about the arrow of time, causality, and the nature of change are also raised by the phenomenon of temporal dilation. Philosophically exploring time dilation's ramifications deepens our comprehension of the multifaceted nature of time, reality, and our place in the cosmos.

Printed in Great Britain
by Amazon